Maximum-Likelihood Deconvolution

Jerry M. Mendel

Maximum-Likelihood Deconvolution

A Journey into Model-Based Signal Processing

C.S. Burrus
Consulting Editor

With 126 Illustrations

Springer-Verlag
New York Berlin Heidelberg
London Paris Tokyo Hong Kong

Jerry M. Mendel
Department of Electrical Engineering-Systems
University of Southern California
Los Angeles, California 90089
U.S.A.

Library of Congress Cataloging-in-Publication Data

Mendel, Jerry M., 1938-
    Maximum-likelihood deconvolution: a journey into model-based
signal processing/Jerry M. Mendel; C.S. Burrus, consulting
editor.
        p. cm.
    Includes bibliographical references.
    ISBN-13:978-1-4612-7985-3 (alk. paper)
    1. Signal processing. 2. Seismic reflection method—
Deconvolution. 3. Estimation theory.   I. Burrus, C.S.
II. Title.
TK5102.5.M376   1990
621.382'2—dc20   89-49051

Printed on acid-free paper.
© 1990 by Springer-Verlag New York Inc.
Softcover reprint of the hardcover 1st edition 1990

Camera-ready copy supplied by the author.

9 8 7 6 5 4 3 2 1

ISBN-13:978-1-4612-7985-3   e-ISBN-13:978-1-4612-3370-1
DOI: 10.1007/978-1-4612-3370-1

To Enders A. Robinson,
father of deconvolution, inspiration, and friend

Preface

Maximum-likelihood deconvolution is a culmination of many years of people's efforts. It can be presented from at least two very different points of view. In most of the journal articles it is couched in the language of state-variable models and estimation theory, both of which , generally speaking, are foreign to many signal processors. This earlier presentation of maximum-likelihood deconvolution was due to this author's own background in control and estimation theory.

This first way of presenting maximum-likelihood deconvolution is unnecessary. The essence of the entire method can be explained much more clearly, and certainly, more simply, using the well-known convolutional model and some relatively simple ideas from optimization theory. Both of these areas should be well known to signal processors. This book adopts this second approach to maximum-likelihood deconvolution. In so doing, I hope that this powerful body of techniques will be readily understood and used by many practicing signal processors.

Although the explanation of maximum-likelihood deconvolution presented herein is in the context of reflection seismology, the theory is also applicable in many other fields, such as nondestructive testing, radar, speech, sonar, and equalization.

In my earlier book on this subject ("Optimal Seismic Deconvolution: an Estimation-Based Approach"), the entire theory was presented for an input sequence (i.e., reflectivity sequence, message sequence) that did not include an additional low-level background noise (i.e., backscatter) term. Only at the very end of the development did I explain how to include this important effect. Because of its importance to the practical implementation of maximum-likelihood deconvolution for real data, I introduce this important effect right at the beginning of this book. In this way, I hope to remove some of the mystery that may have surrounded the way in which this effect was handled in the earlier book.

Readers who are familiar with the first approach described above may be wondering whether all of the many publications which stress the recursive implementations of the maximum-likelihood techniques are now to be discarded. Don't do it, because, as we show in this book, these recursive implementations are the only computationally feasible ones for implementing maximum-likelihood deconvolution. So, whereas it is very straightforward to develop the theory of maximum-likelihood deconvolution using the convolutional model and optimization theory, this approach does not lead to algorithms which are practical for computing purposes. Recursive algorithms are, as we show, *orders of magnitude* faster than the batch algorithms that are associated with the convolutional model.

The book's sub-title "A Journey into Model-Based Signal Processing," reflects the fact that all of the algorithms that are developed within the framework of maximum-likelihood deconvolution are based on a signal model, namely the convolutional model. An example of a signal processing algorithm that is not based on a signal model is the FFT.

A reader who is not too mathematically inclined can still understand the entire theory of maximum-liklihood deconvolution by reading Chapters 1 - 6 and 9. All of the mathematical derivations are collected together in Chapters 7 and 8.

With this book, I hope that maximum-likelihood deconvolution will now be accessible to everyone.

Acknowledgments

I wish to thank John Goutsias, Chong-Yung Chi, A-Chuan Hsueh, and Philip Carrion, for carefully reading the manuscript and making many suggestions for its improvement. Additionally, I would like to thank Catherine Montagna, Georgia Lum, Kris Pendleton, Jill Meschke and John Day, all of whom devoted a great amount of time and effort to the preparation of this book.

Contents

Chapter 4 - Maximizing Likelihood

Chapter 5 - Properties and Performance

Chapter 6 - Examples

Chapter 7 - Mathematical Details for Chapter 4

Chapter 8 - Mathematical Details for Chapter 5

Chapter 9 - Computational Considerations

1
Introduction

1.1 Introduction

Convolution is by far the most important operation that describes the behavior of a linear time-invariant (LTI) dynamical system. It is the operation of convolution that tells us how to compute the output of a LTI system from its input and impulse response (IR), i.e.,

$$\text{output} = \text{input} * \text{IR} \qquad (1\text{-}1)$$

where * denotes the mathematical operation of convolution. Convolution is associated with the "forward problem" of generating the response of a LTI system from known values of its input and IR.

Deconvolution is the unraveling of convolution. It is associated with the "inverse problem " of generating the input to the LTI system from known values of its output and IR. Let IR^{-1} denote the mathematical inverse of a system's impulse response. Because $\text{IR}*\text{IR}^{-1}$ equals a delta function, and the convolution of a delta function with another function (e.g., the input or output) equals that other function,

$$\text{input} = \text{output}*\text{IR}^{-1} \qquad (1\text{-}2)$$

Deconvolution is needed in many branches of science and engineering. In a communication system, such as a telephone, our spoken words (i.e., the message) can be distorted by the telephone system that exists between the mouthpieces at the transmitting and receiving ends. Think of this telephone system as a dynamical system that can be characterized by its impulse response. This IR smears out the message, resulting in *intersymbol interference*. Unless this smearing is undone, the party at the other end of the line will have great difficulty in understanding the spoken message. Deconvolution is used to invert the effect of the telephone system. Of course, this must be done in real time or else there would be unpleasant delays between speaker and listener. Real-time communication deconvolution is known as *equalization*.

In a seismic system the reflectivity that characterizes the behavior of the earth acts as the seismic message. In a reflection seismology experiment (e.g., Robinson and Treitel, 1980) this message is smeared out by the test signal (i.e., seismic source) that probes the earth. The received signal, which is recorded either by a geophone or hydrophone, must be deconvolved in order to undo the effects of the seismic source. Seismic deconvolution does not have to be performed in real time. It is usually carried out, after the completion of a reflection seismology experiment, in a data processing center.

If convolution is the most important LTI system operation, then deconvolution runs a close second. What makes deconvolution so difficult and challenging?

First of all, it is not really possible to obtain the operation IR^{-1} simply by "inverting" a system's IR. If , for example, a system is nonminimum phase (e.g., mixed-delay), so that some of its zeros lie outside of the unit circle in the complex z-domain, then IR^{-1} will be an unstable operation because some of its poles will be unstable, because they will lie outside of the unit circle. So, direct inversion of IR is never recommended.

Additionally, the measured output of a LTI system is often corrupted by additive noise, i.e.,

$$\text{measured output} = \text{input*IR} + \text{noise}, \qquad (1\text{-}3)$$

so that

$$\text{input} = \text{measured output*IR}^{-1} - \text{noise*IR}^{-1} \qquad (1\text{-}4)$$

Because we never know values for the noise we can never compute the input from this formula. Even if we can compute IR^{-1}, we can never compute noise*IR^{-1}. Neglecting this term can lead to serious errors in reconstructing the input.

Finally, because we can never compute the exact inverse operation, it may be very difficult to obtain a high resolution version of the input instead of a blurred version. To understand this, observe that

$$\begin{aligned}
\text{input} &= \text{measured output*IR}^{-1} - \text{noise*IR}^{-1} \\
&= (\text{input*IR} + \text{noise})\text{*IR}^{-1} - \text{noise*IR}^{-1} \\
&= \text{input*}(\text{IR*IR}^{-1}) = \text{input*resolution function} \qquad (1\text{-}5)
\end{aligned}$$

where

$$\text{resolution function} = \text{IR*IR}^{-1} \qquad (1\text{-}6)$$

If IR^{-1} is a perfect inverse operation of IR, then the resolution function is a delta function. No blurring occurs at all in this case, because when resolution function = delta function, input*delta function = input. If IR^{-1} is less than a perfect inverse operation, then $IR*IR^{-1}$ does not equal a delta function. Instead, it equals a smeared out delta function, where the amount of smearing depends on the bandwidth of the IR operation and signal-to-noise ratio (see Figure 1-1). For additional interesting discussions on resolution, see Berkhout (1974).

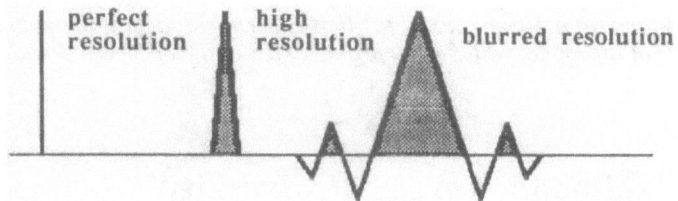

Figure 1-1. Resolution Functions.

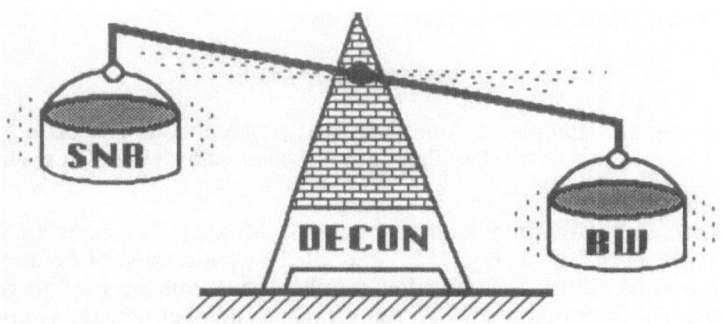

Figure 1-2. The balancing act performed by deconvolution between bandwidth (BW) and signal-to-noise ratio (SNR).

The design of a deconvolution operator requires a careful balancing of bandwidth and signal-to-noise ratio effects (see Figure 1-2). A design procedure that is based on deterministic principles (i.e., a "deterministic" design procedure) is inadequate because it totally neglects the signal-to-noise-ratio effects. A design procedure that is based on stochastic (i.e., random) principles (i.e., a "stochastic" design procedure) handles both effects.

Stochastic design procedures can lead to two types of deconvolution operations, linear or nonlinear. Linear operations generally lead to more smeared out resolution functions than do nonlinear deconvolution operations; hence, high resolution deconvolution requires a design methodology that has the potential for leading to nonlinear deconvolution operations.

1.2   Our Approach

Our approach to deconvolution is based on the widely used and highly regarded maximum-likelihood method, developed circa 1920 by the brilliant British statistician R. A. Fisher (e.g., Fisher, 1922 and 1925). This method, stated in the context of the seismic experiment, is: *given the data from the seismic experiment and an assumed model for that experiment, determine values for the parameters of the model which most probably led to the observed data.*

1.3   Likelihood Versus Probability

I am sure that at one time or another you have used the expressions "This is a very likely ..." and "This is highly probable." These expressions suggest that likelihood and probability are related. In fact

likelihood is proportional to probability.

The following examples demonstrate that probability is associated with a forward experiment whereas likelihood is associated with an inverse experiment.

Example 1-1. [Adapted from: Jerry M. Mendel, "Lessons in Digital Estimation Theory," ©1987, pg. 82. Reprinted by permission of Prentice-Hall, Inc., Englewood Cliffs, NJ.]   Random number generators are used to generate sequences of random numbers. To run a random number generator, you must first specify a probability model and then the parameters of that model. The Gaussian model is often used. Its probability density function is the familiar bell-shaped curve that is completely characterized by two parameters (i.e., numbers) , its mean m and variance v. When we set m and v at specific values (e.g., m = 0 and v = 0.4) we shall think of these values as the *true* values of m and v. We shall denote these true values as m_{TRUE} and v_{TRUE}. Having fixed the values of m and v , we now turn the random generator on. A stream of random numbers is obtained at its output. The first number is denoted n(1), the second n(2), the third n(3), etc. In general, the k th number is denoted n(k). We usually order these numbers in the sequence

$$n(1), \; n(2), \; n(3), \; ... \; , \; n(k), \; ... \; .$$

This is a Gaussian random sequence. Turn the random number generator off and then on again and a completely different sequence (i.e., realization) will be obtained, say

$$n'(1), \; n'(2), \; n'(3), \; ... \; , \; n'(k), ... \; .$$

The Gaussian probability density function for this random number generator is denoted $p(n(k)|m_{n,TRUE}, \; v_{n,TRUE})$, and the numbers we obtain at its output are of course quite dependent on the true parameters $m_{n,TRUE}$ and $v_{n,TRUE}$. The true parameters are sometimes referred to as the *hypothesis* of the probability experiment. The hypothesis must always be fixed in such an experiment.

Example 1-2. [Adapted from: Jerry M. Mendel, "Lessons in Digital Estimation Theory," ©1987, pg. 82. Reprinted by permission of Prentice-Hall, Inc., Englewood Cliffs, NJ.] Suppose we are given a sequence of Gaussian random numbers that are obtained at the output of the Example 1 random number generator, i.e., n(1), n(2), ..., n(N). Now, however, we do not know $m_{n,TRUE}$ and $v_{n,TRUE}$. Is it possible to infer (i.e., "compute" or "estimate") what the values of m_n and v_n are that *most likely* generated the given sequence of numbers? The method of *maximum likelihood* will show us how to do this. The starting point for the determination of m and v will be $p(n(k)| m_n, v_n)$, where now n(k) is given (i.e., it's the given data) and m_n and v_n are treated as the unknowns.

1.4 Maximum Likelihood Method

The *maximum-likelihood method* is based on the relatively simple idea that different probability models generate different samples and that any given sample is more likely to have come from some probability models than from others.

In order to apply the maximum-likelihood method to the design of a deconvolution operation (i.e., filter), we must:

1. Specify a probability model for the measured output;

2. Determine a formula for the likelihood function; and,

3. Maximize the likelihood function.

We shall consider each of these steps in Chapters 2, 3, and 4, respectively.

1.5 Comments

A multitude of methods exist for performing deconvolution; hence, its literature is enormous. Unfortunately, there is no single reference that does adequate justice to all the methods of deconvolution.

Although it is not our purpose to survey all methods of deconvolution, we have included a very extensive bibliography on deconvolution at the end of this book. For the uninitiated, the textbooks, or books of reprints, are excellent places to start on the journey into deconvolution.

This monograph traverses only one of the tributaries of deconvolution, namely the maximum-liklelihood tributary. This tributary has been very well explored. It has a beginning and an end, and has very interesting properties that can be explained using well-established statistical techniques.

2
Convolutional Model

2.1 Introduction

The basic convolutional model is (see Figure 2-1)

$$\text{measured output} = \text{output} + \text{noise} = \text{input*IR} + \text{noise}$$

In this chapter we describe the three components of this model, i.e., input, IR, and noise so that we can compute a formula for the likelihood function. Before doing this, we pause briefly to relate the reflection seismology experiment to the convolutional model.

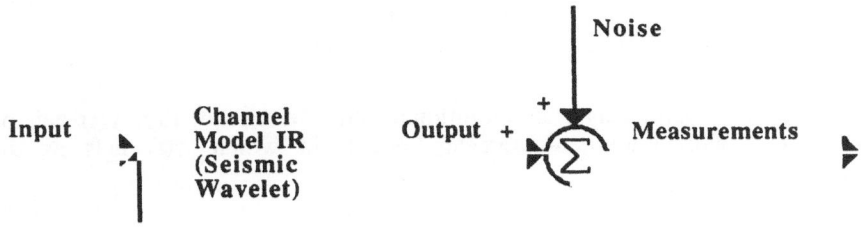

Figure 2-1.   Convolutional model.

2.2 The Seismic Convolutional Model

A layered earth is a distributed parameter system (Mendel, 1986). In the simplest case it is described by a lossless wave equation that is a partial differential equation. Recall that the solution to this wave equation is a time-delayed and scaled replica of the input signal, say w(t). Because a layered earth has interfaces (i.e., boundaries), any internal signal is an infinite series of time-delayed and scaled replicas of w(t) (e.g., Robinson and Treitel, 1980). A ray description can be used to portray the components of such a signal (Figure 2-2).

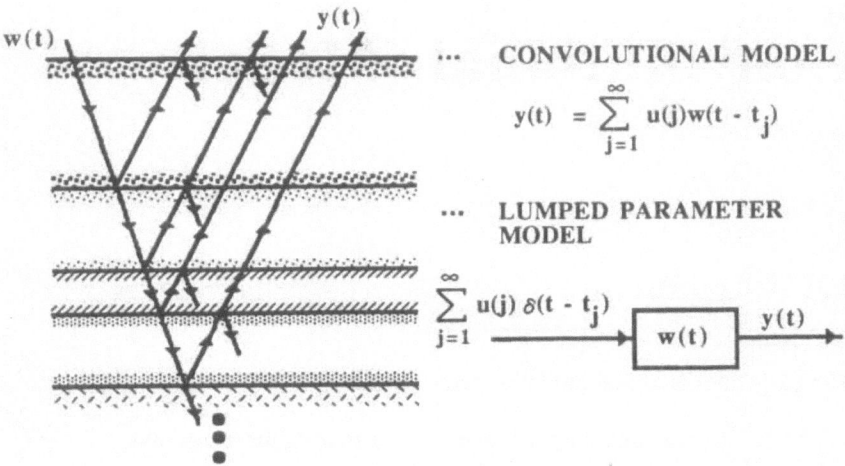

CONVOLUTIONAL MODEL

$$y(t) = \sum_{j=1}^{\infty} u(j)w(t - t_j)$$

... LUMPED PARAMETER MODEL

Figure 2-2. Layered earth interpreted as a convolutional model.

Output signal y(t) is therefore given by the infinite series

$$\sum_{j=1}^{\infty} u(j)w(t - t_j),$$

which we recognize as a convolutional sum. Consequently, y(t) can be associated with a lumped-parameter model, in which w(t) acts as the system's IR and

$$\sum_{j=1}^{\infty} u(j)\delta(t - t_j)$$

acts as the system's input (i.e., the earth's reflectivity sequence). Although this is very useful for signal processing purposes, we have lost the ability to reconstruct the internal states of the layered system. Equivalence between the distributed and lumped parameter systems is only for the output signal y(t).

The lumped parameter model is *counterintuitive* (Mendel, 1983) because earth properties, as characterized by

$$\sum_{j=1}^{\infty} u(j)\delta(t - t_j)$$

excite the seismic source, w(t). From a mathematical point of view, this counterintuitive model is more useful than the one in which w(t) excites an IR given by

$$\sum_{j=1}^{\infty} u(j)\delta(t-t_j),$$

because w(t) can often be modeled by a finite-dimensional model, whereas

$$\sum_{j=1}^{\infty} u(j)\delta(t-t_j)$$

cannot be, and is often assumed to be a random process.

2.3 Input

Our fundamental assumptions about the input signal are that it is *random* and *white*. By white we mean that values of the input signal are completely independent from one value of time to the next. This means that past values of this type of input signal give us no clue about their present or future values, and vice-versa. This whiteness assumption is made in most works on seismic deconvolution, and is sometimes called the "random reflector" hypothesis.

Robinson (1983, pp. 239-241, reprinted by permission of Prentice-Hall, Inc., Englewood Cliffs, NJ) makes the following interesting remarks about the random reflector hypothesis: "A hypothesis is an assertion subject to verification; it is a conjecture that, within a theory or ideational framework, accounts for a set of facts and that can be used as a basis for further investigation. It is a proposition stated as a basis for argument and reasoning. The random reflection hypothesis represents a working proposition. It does not represent absolute truth. The random reflection hypothesis does, however, represent a picture that is supported by the data in enough cases so that it is useful. The random reflection hypothesis is a benchmark as to the statistical distribution of the reflection coefficients used as a reference point in a seismic survey of a completely unknown region.

"... This hypothesis would be expected to hold in areas where the various sediments were laid down over geologic time in a nonpredictable, random manner. As a result, the reflection coefficients would be unrelated to each other and thus would form a random white noise sequence.

"When we say that a reflectivity sequence is random, we do not mean that its autocorrelation for nonzero shifts is identically zero, but that they more or less lie in a band given by the probable range. Some will lie outside this range, of course, but most should lie within it. An actual reflectivity function can be computed from the well logs from an oil well, and then the autocorrelation of the empirical reflectivity function can be computed. We would plot this autocorrelation to see how much it deviates from the theoretical spike autocorrelation. If the deviation is within statistical limits, then we say that the random reflectivity hypothesis is upheld for the geologic sedimentary column represented by that particular oil well. This computation has been done for oil wells in many different oil fields, and the results show that random reflectivity represents a good working hypothesis for the explorationist who is working in an unexplored area. The situation corresponds to card games. A good working hypothesis for card games is that the card shuffle is random. The shuffle mixes the playing cards together to change their order of arrangement. The shuffled deck of cards lies on the table before us, even as the layers of an unknown sedimentary basin appear as random and unknown as cards in a shuffled deck. If we drew the cards one by one and found them black, red, black, red, black, red --- all the way to the last --- we would be surprised, even as we would be surprised if the reflection coefficients were plus, minus, plus, minus, plus, minus --- all the way to the deepest interface.

"In playing with a particular dealer, or in exploring a particular basin, as more experience is gained, certain idiosyncrasies might come out as significant, definite biases. For such a dealer, we would then want to modify our hypothesis. In the same way, as an explorationist learns more about an area, he would want to modify the hypothesis in accord with what he has learned about the actual reflection coefficient distribution present in the sedimentary column.

"In exploration, we are always working against time; time is money. We must start with a noncommittal hypothesis and then modify it only to the extent that time and money will allow. The random reflectivity hypothesis is the best such initial hypothesis."

What if the input signal is random but is not white? Such a signal is then known as *colored noise*. Some recent statistical studies of well logs suggest that a well's reflectivity sequence is colored (Walden and Hoskin, 1986). Colored noise can always be generated by applying white noise to a coloring filter. Suppose we denote the impulse response of such a filter as IR_{CF}. Then

$$\text{colored input sequence} = \text{white input sequence} * IR_{CF} , \qquad (2\text{-}1)$$

so that

$$
\begin{aligned}
\text{measured output} \ &= \text{colored input sequence} * IR + \text{noise} \\
&= \text{white input sequence} * IR_{CF} * IR + \text{noise} \\
&= \text{white input sequence} * IR' + \text{noise} \\
&= \text{input} * IR' + \text{noise} \qquad (2\text{-}2)
\end{aligned}
$$

where

$$IR' = IR_{CF} * IR \qquad (2\text{-}3)$$

Once again our convolutional model can be expressed in terms of a white input sequence that is now applied to a more complicated channel model (i.e., IR' is the cascade of IR_{CF} and IR). Consequently, assuming a white input sequence is not a restrictive assumption.

White sequences come in many guises. The most interesting types for us will be: (a) Gaussian, (b) Bernoulli, (c) Bernoulli-Gaussian, and (d) Bernoulli-Gaussian plus backscatter. All of the early works on deconvolution, such as predictive deconvolution (Robinson, 1967), assume that the reflectivity is a Gaussian white sequence. More recent, high resolution deconvolution methods, such as minimum entropy deconvolution (Wiggins, 1977) assume a nongaussian model.

Other processes that have been recently studied for the reflectivity sequence are the Generalized Gaussian (Gray, 1979) and a mixture of two Laplace distributions (Walden and Hoskin, 1986).

2.3.1 Gaussian White Sequences

Gaussian white sequences were obtained at the output of our Example 1 Gaussian random number generator. An example of a Gaussian sequence is depicted in Figure 2-3. We shall use the symbol u(k) to denote an element of the input sequence. The numbers

$$u(1), u(2), \dots , u(k), \dots , u(N)$$

denote a Gaussian white sequence with N elements. We think of these elements as occuring at time points 1, 2, ... , k, ... , N. This sequence is completely characterized by its mean and variance (Papoulis, 1984). For our work we assume that the mean equals zero; hence, our Gaussian white sequence will be characterized by the single parameter, variance v_u.

Entropy corresponds to the uncertainty about a random signal and also equals the information gained when a signal is observed. A Gaussian sequence is known to be one of maximum entropy, i.e., it is a least-informative signal.

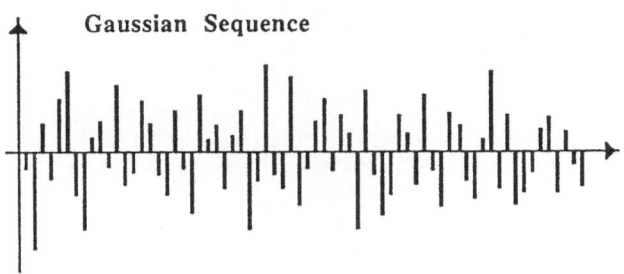

Figure 2-3. A Gaussian white sequence. Values of this sequence are independent from one time point to the next.

2.3.2 Bernoulli White Sequences

A random sequence of zeros and ones is called a Bernoulli sequence. We can think of generating such a sequence by tossing a coin (see Figure 2-4). Everytime a head occurs we call it a one, and everytime a tail occurs we call it a zero. In a "fair" coin tossing experiment the probability of a head occuring equals the probability of a tail occuring. Both probabilities equal 1/2. A Bernoulli sequence can be generated from a fair coin tossing experiment; but, greater variety for such sequences can be obtained from "unfair" coin tossing experiments. If, for example, we want to create a sequence that is very sparse (i.e., lots of zeros and only a small number of ones) we would set the probability of a head occuring equal to a number much smaller than 1/2 (e.g., 0.05), and the probability of a tail occuring equal to a number close to unity (e.g., recall that the probability of a tail plus the probability of a head equals unity, so when probability of a head equals 0.05, the probability of a tail must equal 0.95). An example of a Bernoulli sequence is depicted in Figure 2-5.

We shall use the symbol q(k) to denote an element of the Bernoulli sequence. The numbers

$$q(1), q(2), \dots , q(k), \dots , q(N)$$

denote a Bernoulli sequence with N elements. As in the case of a Gaussian sequence, we think of these elements as occuring at the time points 1, 2, ... , k, ... , N. There are 2^N possible Bernoulli sequences. Each sequence is called a *realization*. Actually, we can choose any one of the 2^N realizations simply by fixing those time points at which we want a one to occur.

Each element of a Bernoulli sequence is described by a probability mass function that is completely characterized by one parameter, λ. This parameter not only equals the mean value of the Bernoulli sequence, but it also equals the

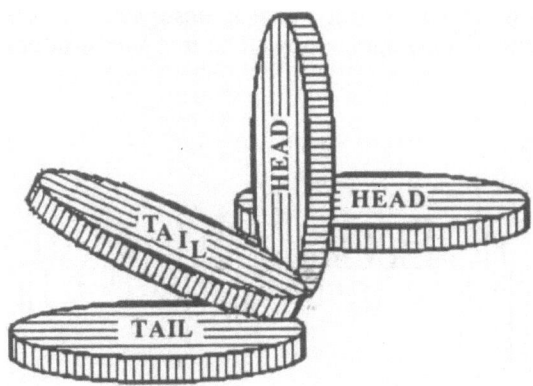

Figure 2-4. Coin tossing experiment.

Bernoulli Sequence

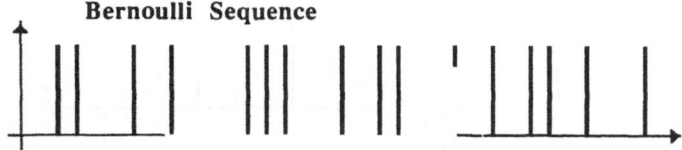

Figure 2-5. A Bernoulli white sequence. The number of spikes in this sequence is controlled by a single parameter, λ.

variance of that sequence (Papoulis, 1984). In general, λ can range from zero to unity, and $\Pr[q(k)] = \lambda$ if $q(k) = 1$ or $\Pr[q(k)] = 1 - \lambda$ if $q(k) = 0$.

The entropy of the Bernoulli distributed event random variable $q(k)$ is $-\lambda\ln(\lambda) - (1 - \lambda)\ln(1 - \lambda)$ (e.g., Mansuripur, 1987). It has a minimum when $\lambda = 0$ or 1 and a maximum when $\lambda = 1/2$. Consequently, the most entropic (least informative) values of the Bernoulli sequence occur when λ is close to 1/2, whereas the least entropic (most informative) values of the Bernoulli sequence occur when λ is close to zero or unity.

2.3.3 Bernoulli-Gaussian White Sequences

Nongaussian input signals play a key role in maximum-likelihood deconvolution, because they lead to nonlinear deconvolution operations. A Bernoulli-Gaussian sequence is one such nongaussian signal. It can be obtained by multiplying the elements of a Bernoulli sequence by the respective elements of a Gaussian sequence. An example of such a sequence is depicted in Figure 2-6. It was obtained by multiplying the Bernoulli sequence in Figure 2-5 by a Gaussian sequence.

Of course, another way to view the creation of a Bernoulli-Gaussian sequence is that at every point in the Bernoulli sequence where a one occurs we turn on our Gaussian random number generator, and replace the one by the output of that generator. When we multiply the Bernoulli sequence by the Gaussian sequence we are performing this replacement not only at the points in the Bernoulli sequence where a one occurs, but at all points. Naturally, we never see the effects of this at the zero value points in the Bernoulli sequence.

We have just described a *product model* for the Bernoulli-Gaussian sequence. Letting $q(k)$ denote an element of the Bernoulli sequence, and $r(k)$ an element of the Gaussian sequence, each element of the Bernoulli-Gaussian sequence equals $r(k)q(k)$. We often refer to $q(k)$ as an *event sequence*, and $r(k)$ as an *amplitude sequence*. Letting $u(k)$ denote the input signal, we have

$$u(k) = r(k)q(k) \qquad (2\text{-}4)$$

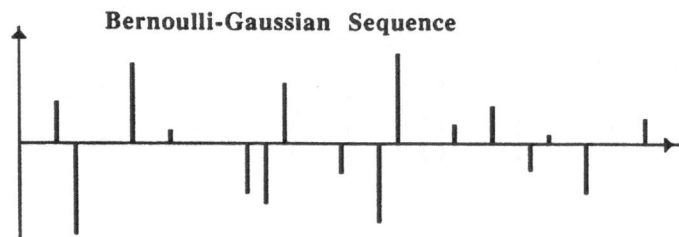

Figure 2-6. A Bernoulli-Gaussian white sequence.

The numbers

$$r(1)q(1), r(2)q(2), \dots, r(k)q(k), \dots, r(N)q(N)$$

denote a Bernoulli-Gaussian sequence with N elements. These elements occur at the time points 1, 2, ... , k, ... , N. The Bernoulli-Gaussian sequence requires two parameters to completely characterize it: λ, which is the probability parameter that is associated with the Bernoulli event sequence, and v_r, which is the variance of the Gaussian amplitude sequence.

2.3.4 Bernoulli-Gaussian plus Backscatter Sequences

In the seismic deconvolution problem, the input signal is the reflectivity function that characterizes the earth. When we model this input just as a Bernoulli-Gaussian sequence, we are accounting for the earth's major layering effects. The earth is also filled with point diffractors which produce a scattering effect. We model this effect by another zero mean white Gaussian sequence $u_B(k)$, which we call a *backscatter sequence*. The Bernoulli-Gaussian plus backscatter input model is

$$u(k) = r(k)q(k) + u_B(k), \tag{2-5}$$

where k = 1, 2, ... , N. An example of such a sequence is depicted in Figure 2-7. It was obtained by adding a Gaussian white sequence to the Bernoulli-Gaussian sequence in Figure 2-6.
    In Chapter 1 we saw that

$$\text{measured output} = \text{input*IR} + \text{noise} \tag{1-3}$$

According to (2-5), the input is now viewed as a sum of two terms, i.e.,

$$\text{input} = \text{earth message} + \text{earth confusion}$$

Bernoulli-Gaussian plus Backscatter

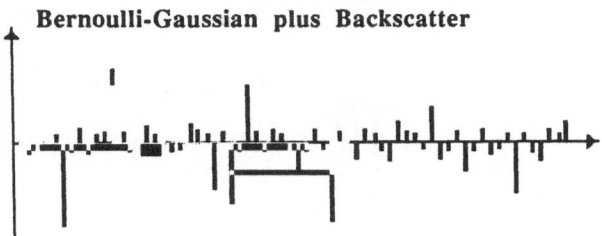

Figure 2-7. A Bernoulli-Gaussian plus backscatter sequence.

The backscatter sequence is used to model the component of the input we refer to as 'earth confusion.' Observe that, unlike the 'noise' term in (1-3), which is additive, backscatter is a convolutional noise effect, i.e., the term earth confusion* IR is convolutional noise.

The Bernoulli-Gaussian plus backscatter input model requires three parameters to completely characterize it: λ, which is the probability parameter that is associated with the Bernoulli event sequence, v_r, which is the variance of the Gaussian amplitude sequence, and v_B, which is the variance of the Gaussian backscatter sequence. This noise model leads to input sequences which often look just like reflectivity sequences that are obtained from well logs; hence, it is a *geologically plausible* model.

Variance v_B is a very interesting parameter. Decreasing it or increasing it permits us to generate input sequences ranging from less entropic to more entropic. By reducing v_B to zero, the Bernoulli-Gaussian plus backscatter model reduces to the less entropic Bernoulli-Gaussian sequence. By increasing v_B we can totally obscure the Bernoulli-Gaussian component. For relatively large values of v_B we obtain a sequence that again looks Gaussian, because the $u_B(k)$ term is then the dominant one in the noise model. This is the most entropic input sequence.

2.4 Channel Model IR (Seismic Wavelet)

We turn now to a description of the channel model's impulse response, i.e., to the seismic wavelet. Seismic wavelets have a frequency content that falls below 200 Hz. They can be rather long in duration, extending as long as 100-200 msec. Generic narrowband and broadband wavelets are depicted in Figures 2-8 and 2-9, respectively. More specific wavelets are depicted in Figures 2-10, 2-11, and 2-12. The Ricker wavelet is representative of the far field signature from a dynamite land source. The water gun and air gun signatures are representative of the far field signatures of two marine sources.

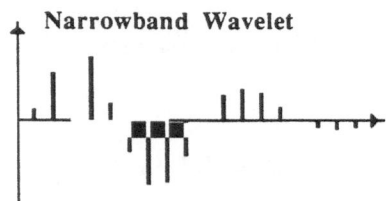

Figure 2-8. A narrowband wavelet.

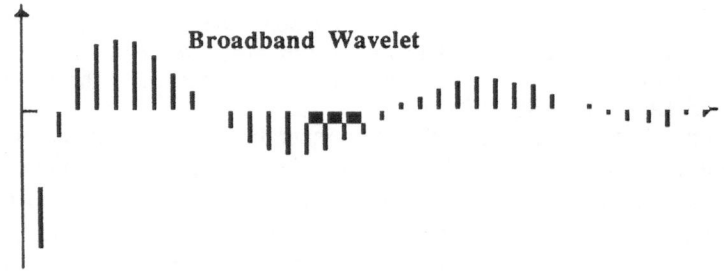

Figure 2-9. A broadband wavelet. Observe the very sharp negative lobe between time points zero and three.

Sampled values of the wavelet are denoted

$$w(0), w(1), w(2), \ldots .$$

Usually, $w(0) = 0$. One model for a wavelet is to use its sampled values directly. Suppose the wavelet is of finite length, i.e., $w(k)$ is of approximately zero value for all values of k greater than M. Then

$$W(z) = w(0) + w(1)z^{-1} + w(2)z^{-2} + \ldots + w(M)z^{-M}, \qquad (2\text{-}6)$$

where $W(z)$ is the z-transform of $w(k)$ and z is the unit advance operator [i.e., $zf(t) = f(t + 1)$]. A model that uses the sampled values of the wavelet directly is called a *moving average (MA)*. The parameters of this model are $w(0), w(1), \ldots , w(M)$. If the wavelet is known to us, then the number M will also be known to us; however, in many important situations the wavelet is unknown to us, so that M is also unknown to us.

There are many other ways to model the wavelet. We often choose to model its z-transform as a ratio of two polynomials, i.e.,

$$W(z) = \frac{b_1 z^{n-1} + b_2 z^{n-2} + \ldots + b_{n-1} z + b_n}{z^n + a_1 z^{n-1} + \ldots + a_{n-1} z + a_n} \qquad (2\text{-}7)$$

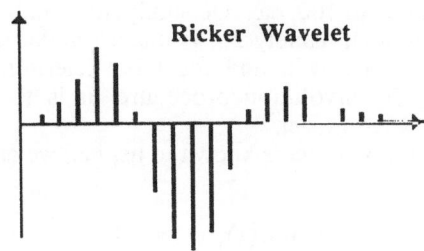

Figure 2-10. A representative Ricker wavelet.

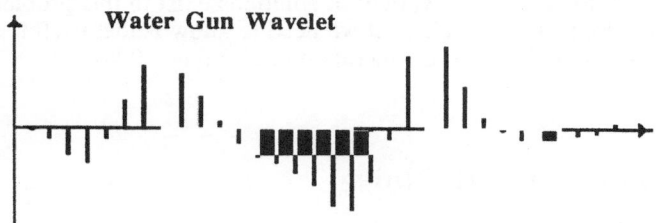

Figure 2-11. A representative water gun wavelet.

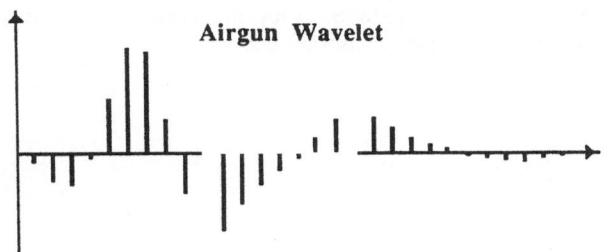

Figure 2-12. A representative airgun wavelet.

This model is called an *autoregressive-moving average (ARMA)*. It is described by the 2n parameters

$$a_1, a_2, \dots, a_n \text{ and } b_1, b_2, \dots, b_n \, .$$

An ARMA wavelet model is known to be *parsimonious* (Ljung, 1987), i.e., it is described by the smallest number of independent parameters. A moving average model is not parsimonious . For seismic wavelets, n, which is the *order* of the wavelet, usually ranges from 3 to 12. For a 1 msec sampling rate, M, the duration of the wavelet, can range from 50 to 200. For a two msec sampling rate, M can range from 25 to 100, etc. Generally, the number of parameters that describe the ARMA model, namely 2n, is much smaller than the number M. When the wavelet is unknown, and must be determined, as part of the maximum-likelihood deconvolution procedure, it is therefore much more expedient to use the ARMA model.

Sometimes a measured wavelet is known to us, i.e., we are given the wavelet samples

$$w(0), w(1), \ldots , w(M).$$

How do we obtain an ARMA model from these values? This is an *approximation problem.* Given the M wavelet samples, we must determine the 2n ARMA parameters, $a_1, a_2, \ldots , a_n, b_1, b_2, \ldots , b_n$, and n such that some measure of the accumulated error between the approximate wavelet and the given M samples is minimized. A variety of solutions exist to this problem, and are beyond the scope of this book. All we need to know is that useful solutions to this problem can be found in the literature (e.g., Kung, 1978).

2.5 Measurement Noise

The third component in our Figure 2-1 convolutional model is the measurement noise, which is assumed to be additive. For simplicity, we shall assume that this noise is zero mean, white, and Gaussian . We shall use the symbol n(k) to denote an element of this sequence. Our measurement noise sequence is characterized by a single parameter, its variance v_n.

2.6 Other Effects

The convolutional model that we have presented neglects many realistic and important effects that are often present in practical applications. These effects, which we list next, can be included in the convolutional model, but at the expense of complexity.

1. *Recording equipment*: The seismogram is measured by either a geophone or a hydrophone. Their signals are transmitted by cable to instruments. The combination of sensor, cable and instrument is referred to as the recording equipment. It is, itself, a dynamical system that often distorts the noise-free seismogram. If the dynamical characteristics of the recording equipment are

known, then they can be included in the convolutional model so that their effects can be removed from the data.

2. *Bandpass filter*: The recorded data are often bandpass-filtered to remove low-frequency surface-wave effects and high-frequency noise effects. A bandpass filter further distorts the recorded signal. If its characteristics are known, its effects can also be included in the convolutional model. Of course, if such filtering has removed a band of frequencies, such as a band of low frequencies, these frequencies can never be restored by means of deconvolution, because they do not exist in the data available at the start of deconvolution.

3. *Colored measurement noise*: Measurement noise, n(k), may not be white. If it has a finite bandwidth, that lies within the bandwidth of the seismic data, then it should be treated as colored noise. As we have mentioned earlier in this chapter, colored noise can be modeled as the output of a coloring filter that is excited by white noise.

4. *Spherical divergence*: The amplitude of a reflected wave is attenuated by a function of depth, due to spherical divergence effects (or, more generally, divergence effects). If this attenuation is not compensated for, it will produce false values for the reflectivity sequence. Compensation is usually achieved by applying a time-varying gain factor to the data. Unless the effects of this time-varying factor are handled properly during deconvolution, erroneous results can occur. Two ways for accomplishing this have been described in the literature, e.g., Kormylo and Mendel (1980) and Mendel (1984). Both require modifying the convolutional model in a very straightforward manner to one that includes time-varying and/or nonstationary effects.

5. *Prespecified horizons*: Occasionally some spike locations are prespecified at a subset of times k^*, perhaps by an interpretor or from well log information. Then $u(k)$ can be expressed as

$$u(k) = q^*(k)r(k) + [1 - q^*(k)]q(k)r(k) \qquad (2\text{-}8)$$

where

$$q^*(k) = \begin{cases} 1, k \in k^* \\ \\ 0, k \notin k^* \end{cases} \qquad (2\text{-}9)$$

From Equations (2-8) and (2-9), observe that

$$u(k) = \begin{cases} r(k), k \in k^* \\ \\ q(k)r(k), k \notin k^* \end{cases} \qquad (2\text{-}10)$$

and, when $k \in k*$ the amplitude of u(k) is still random. It is easy to show, from Equation (2-8), that $E\{u(k)\} = 0$ and

$$v_u(k) = [1 - q*(k)]v_r \lambda + q*(k)v_r \qquad (2\text{-}11)$$

When spike locations are prespecified, u(k) becomes a *nonstationary* white-noise process, because $v_u(k)$ is now a function of k. Mendel (1984) has shown that the whole theory of maximum-likelihood deconvolution can be applied to the case of prespecified horizons, by, for the most part, merely replacing v_u by $v_u(k)$.

6. *Time - varying wavelet*: The seismic wavelet changes its shape as it travels through the earth. This is primarily due to absorption effects, which cause the earth to behave like a bank of low pass filters. We assume that our block of data (i.e., N) is small enough so that the seismic wavelet can be reasonably modeled as time-invariant.

7. *Subjective information*: Sometimes our a priori information is not objective but is subjective. For example, an interpretor may feel that for a given section of data it is best to explain the measurements by a reflectivity which consists of a "relatively small number of significant events"; or, that in any subsequence of data of length N', there should not be "too many events"; or, that in any subsequence of data of length N' there should not be "significant events close together"; or, that a "bright spot" is expected in the data; etc. This subjective knowledge can be quantified using fuzzy set theory to form fuzzy constraints. The fuzzy constraints can be incorporated into the overall deconvolution problem as described in Popoli and Mendel (1989).

2.7 Mathematical Model

Now that we have described the three components of the convolutional model, we collect all of these results together in one place. Figure 2-13 is the same as Figure 2-1, except we now use mathematical symbols to represent the input, channel model IR, output, noise, and measurements. So that we never lose sight of the meaning of these symbols we summarize them:

u(k) = random input signal (reflectivity) which we will model either as a Bernoulli-Gaussian white sequence or as a Bernoulli Gaussian plus backscatter sequence. In the former case, u(k) = r(k)q(k), where r(k) is a white Gaussian amplitude sequence, and q(k) is a Bernoulli event sequence. In the latter case u(k) = r(k)q(k) + u$_B$(k), where u$_B$(k) is a white Gaussian backscatter sequence. Additionally, r(k), q(k) and u$_B$(k) are assumed to be statistically independent of one another.

w(k) =   channel model IR (seismic wavelet) which we will model as an ARMA.

y(k) =   output of the convolutional model; it equals u(k)*w(k).

n(k) =   additive measurement noise, which we will model as white and Gaussian; it is statistically independent of r(k), q(k) and u$_B$(k).

z(k) =   measurements which are available at the output of a physical sensor (e.g., geophone or hydrophone) at the time points 1, 2, ... , N.

The basic convolutional model can now be expressed mathematically, as

$$z(k) = u(k)*w(k) + n(k) , k = 1, 2, ... , N,$$ (2-12)

or, as

$$z(k) = [r(k)q(k) + u_B(k)]*w(k) + n(k), \quad k = 1, 2, ... , N$$ (2-13)

Sometimes it is useful to collect all N measurements together. This is easily done by stacking the N measurements into a vector z, i.e.,

$$z = col(z(1), z(2), ... , z(N)),$$ (2-14)

where col( ) denotes a column vector. Using the fact that

$$u(k)*w(k) = \sum_{j=1}^{k} u(j)w(k-j)$$

we observe that

$$z(1) = w(0)u(1) + n(1)$$
$$z(2) = w(1)u(1) + w(0)u(2) + n(2)$$
$$z(3) = w(2)u(1) + w(1)u(2) + w(0)u(3) + n(3)$$
$$\cdots$$
$$z(N) = w(N-1)u(1) + w(N-2)u(2) + ... + w(0)u(N) + n(N).$$

These N equations can be written in a more compact form, as the following *vector measurement equation:*

$$z = Wu + n \qquad (2\text{-}15)$$

where

$$u = col(u(1), u(2), \dots, u(N)), \qquad (2\text{-}16)$$

$$n = col(n(1), n(2), \dots, n(N)), \qquad (2\text{-}17)$$

and

$$W = \begin{pmatrix} w(0) & 0 & \cdots & 0 \\ w(1) & w(0) & \cdots & 0 \\ \cdot & \cdot & \cdot & \cdot \\ \cdot & \cdot & \cdot & \cdot \\ \cdot & \cdot & \cdot & \cdot \\ w(N-1) & w(N-2) & \cdots & w(0) \end{pmatrix} \qquad (2\text{-}18)$$

If $w(k) \simeq 0$ for $k > M$, then many of the terms in matrix W will be zero; however, there is no loss in generality showing these terms as though they are not zero.

Using the fact that $u(k) = r(k)q(k) + u_B(k)$, we can further express the vector measurement equation, as

$$z = WQr + Wu_B + n \qquad (2\text{-}19)$$

where

$$Q = diag \ [q(1), q(2), \dots, q(N)], \qquad (2\text{-}20)$$

$$r = col \ [r(1), r(2), \dots, r(N)], \qquad (2\text{-}21)$$

and

$$u_B = col \ [u_B(1), u_B(2), \dots, u_B(N)] \qquad (2\text{-}22)$$

Note that in the equation for Q, diag ( ) denotes a diagonal matrix.

If, as is often the case, $w(0) = 0$, then we should ignore the first measurement $z(1)$, because, in this case, $z(1) = n(1)$. When $w(0) = 0$ we redefine z to be $z = col(z(2), z(3), \dots, z(N))$, and then the diagonal elements of matrix W, in Equation (2-18) are all equal to $w(1)$ instead of $w(0)$. We treat this practical situation as a minor change in notation.

This completes the mathematical description of the convolutional model.

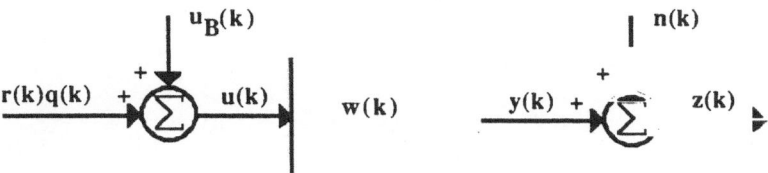

Figure 2-13. Convolutional model revisited.

2.8 Summary

The convolutional model contains three components: input, IR, and noise. The input signal is assumed to be random and white. Gaussian, Bernoulli, Bernoulli-Gaussian, and Bernoulli-Gaussian plus backscatter input models are useful. Most useful in our treatment of seismic deconvolution are the Bernoulli-Gaussian and Bernoulli-Gaussian plus backscatter models. The ARMA wavelet model is most useful because it is parsimonious. Measurement noise is assumed to be additive, white and Gaussian.

The convolutional model is a parametric one. It is completely characterized by parameters that are associated with its three components. Some of these parameters are deterministic, whereas others are random. There are $2n + 4$ deterministic and $3N$ random parameters that describe this model. They are:

> *Deterministic parameters:*
> > *Wavelet parameters:* $\mathbf{a} = \mathrm{col}\,(a_1, a_2, \dots, a_n)$
> > $\mathbf{b} = \mathrm{col}\,(b_1, b_2, \dots, b_n)$
> > *Statistical parameters:* $\mathbf{s} = \mathrm{col}\,(v_r, v_B, v_n, \lambda)$
> *Random parameters:*
> > *Event parameters:* $\mathbf{q} = \mathrm{col}\,[q(1), q(2), \dots, q(N)]$
> > *Amplitude parameters:* $\mathbf{r} = \mathrm{col}\,[r(1), r(2), \dots, r(N)]$
> >
> > *Backscatter parameters:* $\mathbf{u}_B = \mathrm{col}\,[u_B(1), u_B(2), \dots, u_B(N)]$

The reader may now be wondering how it will be possible to determine $2n + 4 + 3N$ parameters from only N measurements. We can do this because we have much more than just the N measurements. We have the *convolutional model* as well as the *probability models* for the 3N random parameters. These models make a solution possible.

Our approach to seismic deconvolution is, therefore, sometimes called *model-based signal processing*.

3
Likelihood

3.1 Introduction

Now that we have established what the unknown parameters are in the deconvolution problem (i.e., $\mathbf{a}$, $\mathbf{b}$, $\mathbf{s}$, $\mathbf{q}$, $\mathbf{r}$, $\mathbf{u}_B$), we create a likelihood function. R. A. Fisher (1922 and 1925) developed the method of maximum likelihood for problems that are characterized just by deterministic parameters. Another method, associated with the name of Thomas Bayes, called the Maximum a Posteriori Method, i.e., MAP (e.g., Sorenson, 1980, and Mendel, 1987a), was developed for problems that are characterized just by random parameters. The known probability models for these random parameters are used in the MAP likelihood function. Our deconvolution problem is a mixture between Fisher and Bayesian likelihood methods, because it contains both deterministic and random parameters (Mendel, 1983). Our approach is to account for both types of parameters in a correct way, i.e., our likelihood function will treat $\mathbf{a}$, $\mathbf{b}$, $\mathbf{s}$, as deterministic, and $\mathbf{r}$, $\mathbf{q}$, $\mathbf{u}_B$, as random. Recall that we do indeed have probability models for $\mathbf{r}$, $\mathbf{q}$, and $\mathbf{u}_B$: $\mathbf{r}$ is multivariate Gaussian, $\mathbf{q}$ is multivariate Bernoulli, and $\mathbf{u}_B$ is multivariate Gaussian. The resulting likelihood function is called an *unconditional likelihood function* (Nahi, 1969), because the random parameters have been properly accounted for. Because the phrase "unconditional likelihood function" is such a mouthful, we shorten it to "likelihood function."

3.2 Loglikelihood

Instead of working with the likelihood function, it is often easier to work with the logarithm of that function, the *loglikelihood function*. This is because the likelihood function contains exponential functions, and taking the natural logarithm of such functions removes their inherent complexities. The reason we are permitted to work with the loglikelihood function instead of the likelihood function is that the logarithm of a function is a monotonic transformation of that function. Whenever the likelihood function is increasing or decreasing, the loglikelihood function is also increasing or decreasing.

Consequently, the point corresponding to the maximum of the likelihood function is also the point corresponding to the maximum of the loglikelihood function.

As pointed out in Chapter 1, likelihood is proportional to probability. Remember that, in probability, all of the parameters in the probability model are fixed, and the measurements are then generated. In likelihood, on the other hand, we are given the measurements and now treat the parameters in the probability model as unknowns.

3.3 Likelihood Function

Regardless of whether we work with likelihood or loglikelihood, what we have established is that in the most general deconvolution problem we are given the measurements, z, and must treat the parameters, a, b, s, q, r, and u_B, as unknowns. Without going into any mathematical details, let us assume that we have been able to arrive at a mathematical formula for the likelihood function. Clearly, it must be a function of a, b, s, q, r, and u_B. Because z is given, we denote this function as

$$L\{a, b, s, q, r, u_B \mid z\}.$$

It is a very complicated function of a, b, s, q, r, and u_B. Maximizing L with respect to all of these unknown parameters is a problem in optimization, a problem which we address in Chapter 4.

3.4 Using Given Information

Whenever additional information is known ahead of time about the components of the convolutional model, we should use it. Depending upon what is known we can obtain a wide range of simpler maximum-likelihood deconvolution (MLD) problems. Here are two specific situations.

1. *Wavelet and statistical parameters are known.* The resulting deconvolution solution is sometimes referred to as *signature deconvolution*. In this case a, b, and s are known, so we suppress the dependence of $L\{\ \}$ on them. Parameters q, r, and u_B must be found by maximizing the function $L\{q, r, u_B \mid z\}$.

2. *Event locations are known, as well as the wavelet and statistical parameters.* In this case q, a, b, and s are known and r and u_B must be found by maximizing the function $L\{r, u_B \mid z\}$. Because r and u_B are both Gaussian, tremendous simplifications occur. In fact, closed form solutions for them can be found. The resulting deconvolution filter is linear.

3.5 Message for the Reader

If you are not mathematically inclined, please skip the next section and proceed to the Summary. You do not need to understand the specific formulas for likelihood function $L\{\ \}$ or loglikelihood function $\mathcal{L}\{\ \}$ [i.e., $\mathcal{L}\{\ \}$ is the natural logarithm of $L\{\ \}$] in order to read Chapter 4.

3.6 Mathematical Likelihood Functions

Using the fact that likelihood is proportional to probability, and assuming a value of unity for the constant of proportionality, we begin by stating that

$$L\{a, b, s, q, r, u_B \mid z\} = p(z, q, r, u_B \mid a, b, s) \tag{3-1}$$

Function $p(z, q, r, u_B \mid a, b, s)$ is the joint probability density function of random vectors z, q, r, and u_B given the deterministic vectors a, b, and s. This likelihood function is conditional with respect to a, b and s and is unconditional with respect to q, r, and u_B.

Here are two elementary conditional probability rules that we will make repeated use of. Let e, f, and g be random variables. Then (e.g., Papoulis, 1984),

$$p(e, f) = p(e \mid f)p(f)$$

and

$$p(e, f, g) = p(e, f \mid g)p(g).$$

Additionally, if e, f, and g are statistically independent, then

$$p(e, f, g) = p(e)p(f)p(g).$$

Applying these simple rules to the preceding likelihood formula, we obtain

$$L\{a, b, s, q, r, u_B \mid z\} = p(z, q, r, u_B \mid a, b, s)$$
$$= p(z \mid q, r, u_B, a, b, s)p(q, r, u_B \mid a,b,s)$$
$$= p(z \mid q, r, u_B, a, b, s) \; p(q \mid a, b, s)$$
$$p(r \mid a, b, s) \; p \; (u_B \mid a, b, s).$$

To arrive at this last expression, we have also used the fact that q, r, and u_B are statistically independent. Next, using the fact that q is a vector of *discrete* random variables, we express $p(q \mid a, b, s)$ as $\Pr (q \mid a, b, s)$; hence,

$$L\{a, b, s, q, r, u_B \mid z\} = p(z \mid q, r, u_B, a, b, s) \Pr (q \mid a, b, s)$$
$$p(r \mid a, b, s) \; p(u_B \mid a, b, s).$$

Because q, r, and u_B do not depend on a and b, we are able to finally express $L\{ \; \}$ as

$$L\{a, b, s, q, r, u_B \mid z\} = p(z \mid q, r, u_B, a, b, s) \Pr (q \mid s)$$
$$p(r \mid s) \; p(u_B \mid s) \tag{3-2}$$

Although this looks like a very formidable expression, it is relatively easy to evaluate each of the four probability functions on its right-hand side. The keys to doing this are the vector measurement equation [see Eq. (2-19)]

$$z = WQr + Wu_B + n, \tag{3-3}$$

the Bernoulli nature of the elements of q, and the Gaussian natures of the elements of r, u_B, and n. From these facts, it is easy to establish that:

$$p(z \mid q, r, u_B, a, b, s) = (2\pi v_n)^{-N/2} \exp \{- (z - WQr - Wu_B)'$$
$$(z - WQr - Wu_B)/2v_n\} ; \tag{3-4}$$

$$\Pr (q \mid s) = \prod_{k=1}^{N} \Pr [q(k) \mid \lambda] = \lambda^{m(q)} (1 - \lambda)^{[N - m(q)]} , \tag{3-5}$$

in which

$$m(q) = \sum_{k=1}^{N} q(k); \tag{3-6}$$

$$p(r \mid s) = (2\pi v_r)^{-N/2} \exp (- r'r/2v_r) ; \tag{3-7}$$

and,

$$p\ (u_B\mid s) = (2\pi v_B)^{-N/2}\ \exp\ (-\ u_B{}'u_B/2v_B) \qquad (3\text{-}8)$$

Putting all of this together, we obtain the following final expression for $L\{\ \}$:

$$
\begin{aligned}
L\{\mathbf{a}, \mathbf{b}, \mathbf{s}, \mathbf{q}, \mathbf{r}, u_B\mid z\} = &\ (2\pi)^{-3N/2}(v_n v_r v_B)^{-N/2}\ \exp\ \{-r'r/2v_r \\
&- (z - \mathbf{W}\mathbf{Q}\mathbf{r} - \mathbf{W}u_B)'(z - \mathbf{W}\mathbf{Q}\mathbf{r} - \mathbf{W}u_B)/2v_n \\
&- u_B{}'u_B/2v_B\ \}\ \lambda^{m(q)}\ (1 - \lambda)^{[N - m(q)]} \qquad (3\text{-}9)
\end{aligned}
$$

3.7 Mathematical Loglikelihood Functions

Because of the strong exponential dependence on the right-hand side of this last expression, it is often easier to work with the loglikelihood function, which is obtained by taking the natural logarithm of $L\{\ \}$. We denote the loglikelihood function $\mathscr{L}\{\ \}$; it is given as

$$
\begin{aligned}
\mathscr{L}\{\mathbf{a}, \mathbf{b}, \mathbf{s}, \mathbf{q}, \mathbf{r}, u_B\mid z\} = &\ -\tfrac{3}{2}N\ln 2\pi - \tfrac{N}{2}\ \ln(v_n v_r v_B)\ r'r/2v_B \\
&- (z - \mathbf{W}\mathbf{Q}\mathbf{r} - \mathbf{W}u_B)'(z - \mathbf{W}\mathbf{Q}\mathbf{r} - \mathbf{W}u_B)/2v_n \\
&- u_B{}'u_B/2v_B + m(q)\ln(\lambda) \\
&+ [N - m(q)]\ln(1 - \lambda) \qquad (3\text{-}10)
\end{aligned}
$$

By convention, we drop the constant term, $-\tfrac{3}{2}N\ln 2\pi$, from this expression, because it does not depend on any of the parameters.

Examining the resulting expression, note that the dependence of $L\{\ \}$ on: $\mathbf{a}$ and $\mathbf{b}$ occurs only in its third term (because $\mathbf{W}$ depends on $\mathbf{a}$ and $\mathbf{b}$); $\mathbf{s}$ occurs in all the terms; $\mathbf{q}$ occurs in the third, fifth, and sixth terms; $\mathbf{r}$ occurs in the second and third terms; and, u_B occurs in the third and fourth terms.

Let us see what happens to $\mathscr{L}\{\ \}$ in the two special cases that were described above.

1. *Wavelet and statistical parameters are known.* In this case $\mathscr{L}\{\ \}$ depends only upon $\mathbf{q}, \mathbf{r},$ and u_B, i.e., $L\{\mathbf{a}, \mathbf{b}, \mathbf{s}, \mathbf{q}, \mathbf{r}, u_B\mid z\} = L\{\mathbf{q}, \mathbf{r}, u_B\mid z\}$, and

$$
\begin{aligned}
\mathscr{L}\{\mathbf{q}, \mathbf{r}, u_B\mid z\} = &\ -r'r/2v_r - u_B{}'u_B/2v_B \\
&- (z - \mathbf{W}\mathbf{Q}\mathbf{r} - \mathbf{W}u_B)'(z - \mathbf{W}\mathbf{Q}\mathbf{r} - \mathbf{W}u_B)/2v_n \\
&+ m(q)\ln(\lambda) + [N - m(q)]\ln(1 - \lambda)
\end{aligned}
$$

Even though this formula looks very similar to the previous one, it is less complicated to maximize because it only depends on $\mathbf{q}, \mathbf{r},$ and u_B.

2. *Event locations are known, as well as the wavelet and statistical parameters.* In this case $\mathscr{L}\{\ \}$ depends only on $\mathbf{r}$ and $\mathbf{u_B}$, i.e., $\mathscr{L}\{\mathbf{a}, \mathbf{b}, \mathbf{s}, \mathbf{q}, \mathbf{r}, \mathbf{u_B} \mid \mathbf{z}\} = \mathscr{L}\{\mathbf{r}, \mathbf{u_B} \mid \mathbf{z}\}$, and

$$\mathscr{L}\{\mathbf{r}, \mathbf{u_B} \mid \mathbf{z}\} = -\,\mathbf{r'r}/2v_r - \mathbf{u_B'u_B}/2v_B$$
$$-\,(\mathbf{z} - \mathbf{WQr} - \mathbf{Wu_B})'(\mathbf{z} - \mathbf{WQr} - \mathbf{Wu_B})/2v_n$$

This formula is definitely simpler looking than the original one. Observe the almost symmetrical roles played by $\mathbf{r}$ and $\mathbf{u_B}$, and that $\mathscr{L}\{\ \}$ is a quadratic function of each of these vectors. It will be a straightforward exercise in vector calculus to maximize $\mathscr{L}\{\ \}$ with respect to $\mathbf{r}$ and $\mathbf{u_B}$. The resulting deconvolution filter will depend linearly on the measurements in $\mathbf{z}$, and is known as a *Minimum-Variance Deconvolution* (MVD) filter. Note, also, that both sources of randomness in this special case are Gaussian. This is another reason why the resulting deconvolution filters are linear.

3.8 Summary

Our deconvolution problem is a mixture between Fisher and Bayesian likelihood methods; it contains both deterministic and random parameters. Our likelihood function is called an unconditional likelihood function, because the random parameters have been properly accounted for. Because of the exponential nature of many likelihood functions it is often easier to work with the loglikelihood function. Maximizing the loglikelihood function is equivalent to maximizing the likelihood function. In either case, we are faced with an optimization problem, a problem we address in the next chapter.

4
Maximizing Likelihood

4.1 Introduction

We have shown that the deconvolution problem can be viewed as an optimization problem. The maximum-likelihood (ML) values of $\mathbf{a}, \mathbf{b}, \mathbf{s}, \mathbf{q}, \mathbf{r}$, and $\mathbf{u}_B$ are the values where either the likelihood function $L\{\mathbf{a}, \mathbf{b}, \mathbf{s}, \mathbf{q}, \mathbf{r}, \mathbf{u}_B \mid \mathbf{z}\}$ or the loglikelihood function $\mathcal{L}\{\mathbf{a}, \mathbf{b}, \mathbf{s}, \mathbf{q}, \mathbf{r}, \mathbf{u}_B \mid \mathbf{z}\}$ attains its maximum. Because of the exponential nature of our likelihood function, we shall focus on maximizing $\mathcal{L}\{\ \}$. Maximum-likelihood values of the parameter vectors are denoted with a superscript ML, e.g., $\mathbf{a}^{ML}, \mathbf{u}_B{}^{ML}$. There are many different methods one can use to maximize $\mathcal{L}\{\ \}$. We shall examine some of these in this chapter. First, however, we must be convinced that maximizing $\mathcal{L}\{\ \}$ is a meaningful thing to do.

4.2 A Rationale

In Chapter 1 we noted that in the ML method we are given the data from a seismic experiment and an assumed model for that experiment, and we then determine values for the parameters of the model which most probably led to the observed data. In essence, *likelihood can be used as the basis for deciding between competing alternative solutions.* Consider the following example.

We are given the seismic trace depicted in Figure 4-1. There is no backscatter. Only the noise parameters in $\mathbf{s}$, namely v_r, v_n, and λ, are known. Because the convolutional model contains two unknowns, namely the wavelet, $w(k)$, and reflectivity, $u(k)$, an infinite number of solutions exist for these quantities that are consistent with the data. Two such solutions are depicted in Figures 4-2 and 4-3. How do we know which one of these solutions is better than the other? We can evaluate the likelihood functions for each solution and decide that the better solution is the one with the largest likelihood function.

In this case, because the noise parameters are known, and there is no backscatter, $\mathscr{L}\{\ \}$ can be decomposed into a sum of three terms (see the footnote to Table 4-1). The first term provides a measure of the size of the reflectivity; the second term provides a measure of the number of spikes in the reflectivity sequence; and, the third term provides a measure of the error between z and its model. The third term is known either as the *residual noise* or the *prediction error*. Table 4-1 summarizes the three components of the loglikelihood function for the Figures 4-2 and 4-3 deconvolution solutions. The results in Figure 4-2 are now judged to be better than those in Figure 4-3 because the loglikelihood value of -98.19 is much larger than its competing value of -158.98. Likelihood has indeed provided us with the rationale for deciding between these two competing solutions. Of course, in reality there are many more than two solutions. Our objective is to find the very best one.

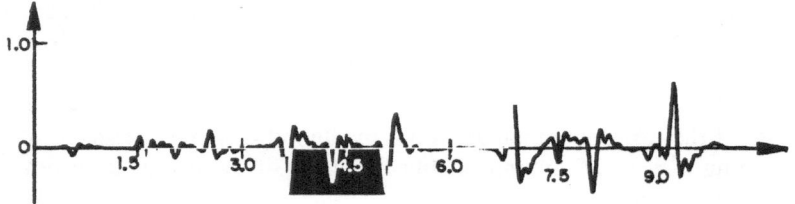

Figure 4-1. A seismic trace (Hampson and Russell, 1983, Fig. 2).

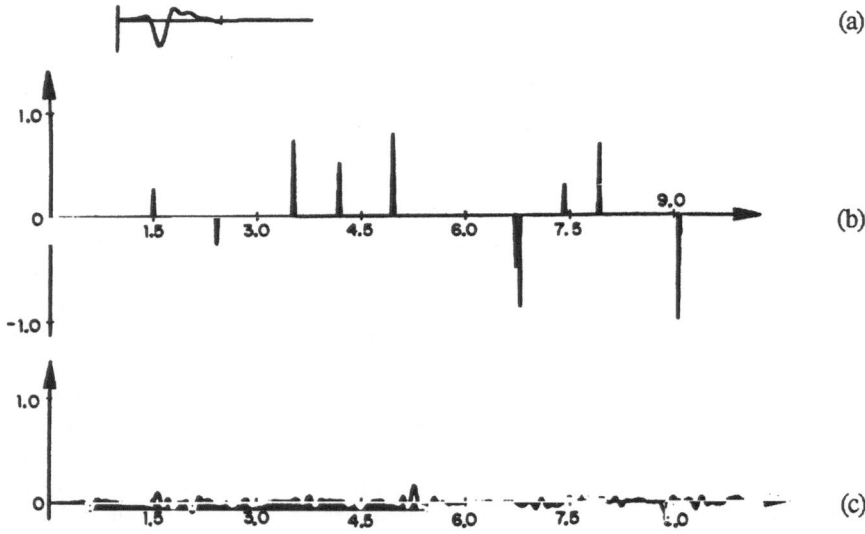

Figure 4-2. Deconvolution solution number 1: (a) wavelet, (b) reflectivity, and (c) residual noise (Hampson and Russell, 1983, Fig. 2).

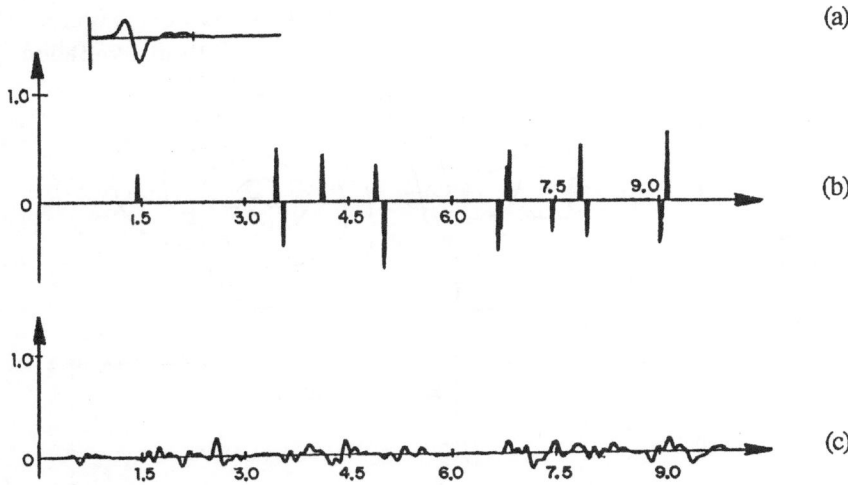

Figure 4-3. Deconvolution solution number 2: (a) wavelet, (b) reflectivity, and (c) residual noise (Hampson and Russell, 1983, Fig. 3).

4.3 Block Component Search Algorithms

Two factors that greatly complicate maximization of the loglikelihood function are its large number of parameters (i.e., $3N + 2n + 4$), and the discrete nature of the N parameters in **q**. Application of popular gradient search algorithms to $\mathcal{L}\{\ \}$ need some modification, because of the parameters in **q**. Derivatives of $\mathcal{L}\{\ \}$ with respect to these zero/one parameters do not exist.

One method for maximizing a function f(x,y) of two (generic) variables is *recursive* (e.g., Goutsias and Mendel, 1986). First, we keep x constant and maximize f(x,y) with respect to y. Then, we keep y constant and maximize f(x,y) with respect to x. These steps are then repeated recursively. Figure 4-4 depicts the steps of this method, which we shall refer to as a *recursive block optimization method*. Let x_i, y_i, denote the values of x and y at the i th iteration of this method, and x_{i+1}, y_{i+1} denote the values of x and y at the i+1 st iteration. For this method to be successful, we require

$$f(x_i, y_i) \le f(x_{i+1}, y_i) \le f(x_{i+1}, y_{i+1}) \ .$$

Our description of this method applies equally as well to functions of more than two variables.

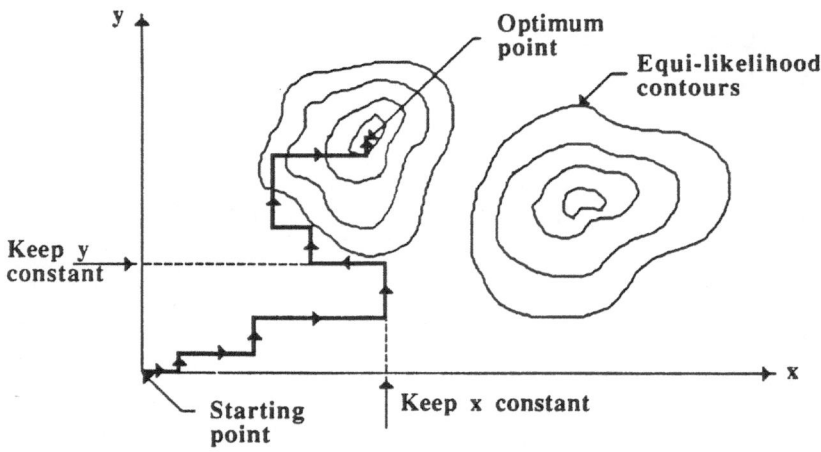

Figure 4-4. Recursive block optimization over a two-dimensional space (Goutsias and Mendel, 1986, Fig. 6).

Table 4-1. Loglikelihood values.

Result	Spike Size	Spike Density	Prediction Error	$\mathscr{L}\{\ \}$*
Figure 4-2	-9.19	-50.00	-39.00	-98.19
Figure 4-3	-6.38	-70.85	-81.75	-158.98

*For the mathematically inclined, $\mathscr{L}\{a, b, q, r, | z\} = -r'r/2v_r + \{m(q)\ln(\lambda) + [N - m(q)]\ln(1-\lambda)\} - (z - WQr)'(z - WQr)/2v_n$. See Chapter 3 [i.e., Eq. (3-10)] for the derivation of this formula.

Here is one way in which this recursive block optimization method can be used to maximize $\mathscr{L}\{\ \}$. Fix the generic variables, x and y, as follows:

$$x = \text{elements of } q, \quad \text{and} \quad y = \text{elements of } a, b, s, r, u_B.$$

Keeping a, b, s, r, and u_B constant and maximizing $\mathscr{L}\{\ \}$ with respect to x corresponds to optimization over a discrete set of variables. A total number of 2^N possible q vectors have to be generated, and the 2^N possible values of $\mathscr{L}\{\ \}$ then have to be computed. The number 2^N is enormous even for modest values of N (e.g., 200); hence, the computational time for doing this is prohibitive, so that some other approach must be found. Detection algorithms provide us with just such an approach. They do not suffer from the computational drawbacks of the direct approach. Detection algorithms are described later in this chapter.

Keeping q constant, and maximizing $\mathcal{L}\{\ \}$ with respect to y can be accomplished in many different ways. Because a, b, s, r, and u_B are continuous in nature, we could use a gradient algorithm to update the entire collection of these y-parameters in one shot. This could still be a very costly thing to do, because there are $2N + 2n + 4$ elements in y.

Figure 4-5 summarizes this two step approach.   We refer to it as a *Block Component Search Algorithm.* Its first step guarantees

$$\mathcal{L}\{a_i, b_i, s_i, q_i, r_i, u_{B,i} \mid z\} \leq \mathcal{L}\{a_i, b_i, s_i, q_{i+1}, r_i, u_{B,i} \mid z\} ,$$

and the second step guarantees

$$\mathcal{L}\{a_i, b_i, s_i, q_{i+1}, r_i, u_{B,i} \mid z\} \leq \mathcal{L}\{a_{i+1}, b_{i+1}, s_{i+1}, q_{i+1}, r_{i+1}, u_{B,i+1} \mid z\} .$$

Combining these two expressions, we see that

$$\mathcal{L}\{a_i, b_i, s_i, q_i, r_i, u_{B,i} \mid z\} \leq \mathcal{L}\{a_i, b_i, s_i, q_{i+1}, r_i, u_{B,i} \mid z\}$$
$$\leq \mathcal{L}\{a_{i+1}, b_{i+1}, s_{i+1}, q_{i+1}, r_{i+1}, u_{B,i+1} \mid z\},$$

which demonstrates that our Block Component Search Algorithm is indeed increasing the loglikelihood function from one iteration to the next. *A Block Component Search Algorithm should achieve a local maximum of $\mathcal{L}$; there is no guarantee that it will achieve a global maximum of $\mathcal{L}$.*

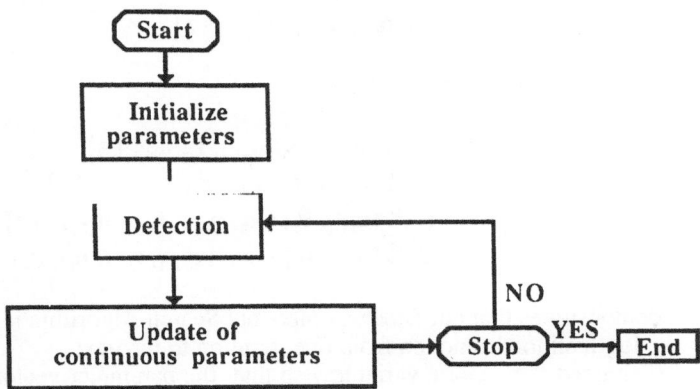

Figure 4-5. Block Component Search Algorithm. Sub-iterations (i.e., additional iterations) may occur within the 'detection' and 'update of continuous parameters' blocks.

A Block Component Search Algorithm does not have to be limited to a two-step procedure. In our case, because generic variable y contains parameters from three distinctly different parts of the convolutional model, it is very useful to further decompose y, into

$$y_1 = \text{elements of } \mathbf{r}, \mathbf{u_B}$$
$$y_2 = \text{elements of } \mathbf{a}, \mathbf{b}$$
$$y_3 = \text{elements of } \mathbf{s}.$$

By this decomposition, y_1 is associated with the random parameters, y_2 with the wavelet's parameters, and y_3 with the remaining statistical parameters.

Figure 4-6 summarizes a four step Block Component Search Algorithm. Its first step guarantees

$$\mathscr{L}\{\mathbf{a}_i, \mathbf{b}_i, \mathbf{s}_i, \mathbf{q}_i, \mathbf{r}_i, \mathbf{u}_{B,i} \mid \mathbf{z}\} \leq \mathscr{L}\{\mathbf{a}_i, \mathbf{b}_i, \mathbf{s}_i, \mathbf{q}_{i+1}, \mathbf{r}_i, \mathbf{u}_{B,i} \mid \mathbf{z}\};$$

its second step guarantees

$$\mathscr{L}\{\mathbf{a}_i, \mathbf{b}_i, \mathbf{s}_i, \mathbf{q}_{i+1}, \mathbf{r}_i, \mathbf{u}_{B,i} \mid \mathbf{z}\} \leq L\{\mathbf{a}_i, \mathbf{b}_i, \mathbf{s}_i, \mathbf{q}_{i+1}, \mathbf{r}_{i+1}, \mathbf{u}_{B,i+1} \mid \mathbf{z}\};$$

its third step guarantees

$$\mathscr{L}\{\mathbf{a}_i, \mathbf{b}_i, \mathbf{s}_i, \mathbf{q}_{i+1}, \mathbf{r}_{i+1}, \mathbf{u}_{B,i+1} \mid \mathbf{z}\} \leq$$
$$\mathscr{L}\{\mathbf{a}_{i+1}, \mathbf{b}_{i+1}, \mathbf{s}_i, \mathbf{q}_{i+1}, \mathbf{r}_{i+1}, \mathbf{u}_{B,i+1} \mid \mathbf{z}\};$$

and, its fourth step guarantees

$$\mathscr{L}\{\mathbf{a}_{i+1}, \mathbf{b}_{i+1}, \mathbf{s}_i, \mathbf{q}_{i+1}, \mathbf{r}_{i+1}, \mathbf{u}_{B,i+1} \mid \mathbf{z}\} \leq$$
$$\mathscr{L}\{\mathbf{a}_{i+1}, \mathbf{b}_{i+1}, \mathbf{s}_{i+1}, \mathbf{q}_{i+1}, \mathbf{r}_{i+1}, \mathbf{u}_{B,i+1} \mid \mathbf{z}\}.$$

Combining these expressions, we see that

$$\mathscr{L}\{\mathbf{a}_i, \mathbf{b}_i, \mathbf{s}_i, \mathbf{q}_i, \mathbf{r}_i, \mathbf{u}_{B,i} \mid \mathbf{z}\} \leq \mathscr{L}\{\mathbf{a}_i, \mathbf{b}_i, \mathbf{s}_i, \mathbf{q}_{i+1}, \mathbf{r}_i, \mathbf{u}_{B,i} \mid \mathbf{z}\}$$
$$\leq \mathscr{L}\{\mathbf{a}_i, \mathbf{b}_i, \mathbf{s}_i, \mathbf{q}_{i+1}, \mathbf{r}_{i+1}, \mathbf{u}_{B,i+1} \mid \mathbf{z}\}$$
$$\leq \mathscr{L}\{\mathbf{a}_{i+1}, \mathbf{b}_{i+1}, \mathbf{s}_i, \mathbf{q}_{i+1}, \mathbf{r}_{i+1}, \mathbf{u}_{B,i+1} \mid \mathbf{z}\}$$
$$\leq \mathscr{L}\{\mathbf{a}_{i+1}, \mathbf{b}_{i+1}, \mathbf{s}_{i+1}, \mathbf{q}_{i+1}, \mathbf{r}_{i+1}, \mathbf{u}_{B,i+1} \mid \mathbf{z}\},$$

which again demonstrates that our Block Component Search Algorithm is indeed increasing the loglikelihood function from one iteration to the next.

We have partitioned the generic variable y so that the parameter vectors $\mathbf{a}$, $\mathbf{b}$, $\mathbf{s}$, $\mathbf{r}$, and $\mathbf{u_B}$ are ordered in a certain way. This led to a Block Component Search Algorithm in which we updated the random parameters first, wavelet parameters second, and statistical parameters third. We can permute the three generic y_1, y_2, and y_3 variables to arrive at 6 different Block Component Search Algorithms. Unless theory dictates otherwise, there is no reason to believe that any one of

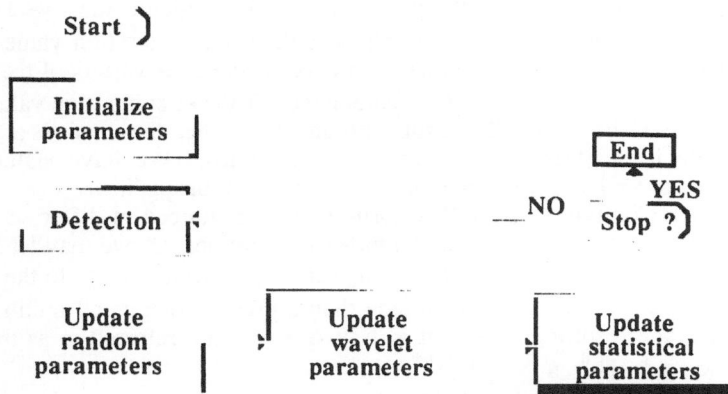

Figure 4-6. Four step Block Component Search Algorithm. Sub-iterations may occur within each major block.

these algorithms is better than any other (although there could be computational timing problems with some of them, and some may lead to faster convergence than others).

Not only can we permute the three generic y_1, y_2, and y_3 variables, but we can also permute the generic x variable with them, so that there are 24 different Block Component Search Algorithms, which are summarized in Table 4-2. There is no apparent advantage for any one of these algorithms over any other.

Table 4-2. 24 Block Component Search Algorithms.

x y_1 y_2 y_3	y_1 x y_2 y_3	y_2 x y_1 y_3	y_3 x y_1 y_2
x y_1 y_3 y_2	y_1 x y_3 y_2	y_2 x y_3 y_1	y_3 x y_2 y_1
x y_2 y_1 y_3	y_1 y_2 x y_3	y_2 y_1 x y_3	y_3 y_1 x y_2
x y_2 y_3 y_1	y_1 y_2 y_3 x	y_2 y_1 y_3 x	y_3 y_1 y_2 x
x y_3 y_1 y_2	y_1 y_3 x y_2	y_2 y_3 x y_1	y_3 y_2 x y_1
x y_3 y_2 y_1	y_1 y_3 y_2 x	y_2 y_3 y_1 x	y_3 y_2 y_1 x

A Block Component Search Algorithm provides us with an optimization strategy. Within each block we are free to choose a specific optimization algorithm, and, sub-iterations may occur within each major block. For example, a mathematical programming algorithm such as steepest descent, conjugate gradient, or Marquardt-Levenberg, can be used to update the y_2 and y_3 parameters from their i th iteration values to their i+1 st iteration values. An MVD algorithm can be used to update the y_1 parameters. A variety of recursive detectors can be used to update the x parameters. The i+1 st iteration values are accepted only if the loglikelihood function increases when these values are used to compute it in place of the i th iteration values. We have stated this replacement procedure above, in terms of a sequence of inequalities.

The Block Component Search Algorithm can be stopped in different ways. We may choose to run it for a fixed number of iterations; or, we may let it stop automatically, by iterating until the change in $\mathcal{L}\{\ \}$, from the i th to the i+1 st iteration becomes smaller in magnitude than a prespecified small number. In either case, the resulting values of $\mathbf{a}$, $\mathbf{b}$, $\mathbf{s}$, $\mathbf{q}$, $\mathbf{r}$, $\mathbf{u}_B$, are referred to as the ML values, i.e., $\mathbf{a}^{ML}$, $\mathbf{b}^{ML}$, $\mathbf{s}^{ML}$, $\mathbf{q}^{ML}$, $\mathbf{r}^{ML}$, $\mathbf{u}_B{}^{ML}$.

The Block Component Search Algorithm is straightforward, but it is also a bit of a brute force approach. Following is a mathematical fact that lets us begin by removing the dependences of $\mathcal{L}\{\ \}$ on $\mathbf{r}$ and $\mathbf{u}_B$. In this way, we are able to reduce the number of unknown parameters by 2N, from 3N + 2n + 4 to N + 2n + 4. Because N can be a large number, such a reduction is very substantial.

4.4 Mathematical Fact

Once again, consider the problem of maximizing a function f(x,y) of the two generic variables x and y. The point $(\hat{x},\hat{y})$, which maximizes f(x,y) can be found from the following four step procedure (Goutsias and Mendel, 1986):

1. Set the partial derivative of f(x,y) with respect to x equal to zero [if f(x,y) is not differentiable with respect to x, then set the partial difference of f(x,y) with respect to x equal to zero]. For every y, call the solution x, that optimizes f(x,y), h(y), i.e., x = h(y). If this solution cannot be obtained, then STOP.

2. Substitute x = h(y) into f(x,y) and call the resulting function g(y), i.e., g(y) = f[h(y), y].

3. Find the point $\hat{y}$ that maximizes g(y).

4. Compute $\hat{x}$ as $\hat{x} = h(\hat{y})$.

A proof of this procedure is given later in Chapter 7. It does not use any derivatives; hence, this procedure can also be applied to a function f(x,y) whose variables can be either continuous or discrete. This procedure can also be generalized to a function of more than two variables.

4.5   Separation Principle

Instead of finding $\mathbf{a}^{ML}$, $\mathbf{b}^{ML}$, $\mathbf{s}^{ML}$, $\mathbf{q}^{ML}$, $\mathbf{r}^{ML}$, and $\mathbf{u_B}^{ML}$ directly by maximizing the loglikelihood function $\mathcal{L}\{\mathbf{a}, \mathbf{b}, \mathbf{s}, \mathbf{q}, \mathbf{r}, \mathbf{u_B} \mid \mathbf{z}\}$, we can first find $\mathbf{a}^{ML}$, $\mathbf{b}^{ML}$, $\mathbf{s}^{ML}$, and $\mathbf{q}^{ML}$ by maximizing a related objective function $\mathcal{M}\{\mathbf{a}, \mathbf{b}, \mathbf{s}, \mathbf{q} \mid \mathbf{z}\}$, where (Kormylo, 1979, Kormylo and Mendel, 1982, or Mendel, 1983)

$$\mathcal{M}\{\mathbf{a}, \mathbf{b}, \mathbf{s}, \mathbf{q} \mid \mathbf{z}\} = -\frac{N}{2} \ln(v_n v_r v_B) - \frac{1}{2} \mathbf{z}'\Omega^{-1}\mathbf{z}$$
$$+ m(\mathbf{q})\ln(\lambda) + [N - m(\mathbf{q})]\ln(1 - \lambda) \qquad (4\text{-}1)$$

[in which Ω, the covariance matrix of measurement vector $\mathbf{z}$ conditioned on knowing the event vector $\mathbf{q}$, is defined in Eq. (7-17)] after which $\mathbf{r}^{ML}$ and $\mathbf{u_B}^{ML}$ are easily computed as linear operations on the data $\mathbf{z}$, i.e.,

$$\mathbf{r}^{ML} = \mathbf{A}(\mathbf{a}^{ML}, \mathbf{b}^{ML}, \mathbf{s}^{ML}, \mathbf{q}^{ML})\, \mathbf{z} \qquad (4\text{-}2)$$

and

$$\mathbf{u_B}^{ML} = \mathbf{B}(\mathbf{a}^{ML}, \mathbf{b}^{ML}, \mathbf{s}^{ML}, \mathbf{q}^{ML})\, \mathbf{z}. \;\square \qquad (4\text{-}3)$$

In order to prove this result we use our *Mathematical Fact* in which we set

$$x = \text{elements of } \mathbf{r} \text{ and } \mathbf{u_B}$$

and

$$y = \text{elements of } \mathbf{a}, \mathbf{b}, \mathbf{s}, \text{ and } \mathbf{q}.$$

The exact formulas for $\mathcal{M}\{\mathbf{a}, \mathbf{b}, \mathbf{s}, \mathbf{q} \mid \mathbf{z}\}$, $\mathbf{A}(\mathbf{a}^{ML}, \mathbf{b}^{ML}, \mathbf{s}^{ML}, \mathbf{q}^{ML})$, and $\mathbf{B}(\mathbf{a}^{ML}, \mathbf{b}^{ML}, \mathbf{s}^{ML}, \mathbf{q}^{ML})$, as well as a proof of the Separation Principle, are given in Chapter 7.

We call this a *Separation Principle* because we determine $\mathbf{r}^{ML}$ and $\mathbf{u_B}^{ML}$ separately from the determinations of $\mathbf{a}^{ML}$, $\mathbf{b}^{ML}$, $\mathbf{s}^{ML}$, and $\mathbf{q}^{ML}$. It is important to understand that this separation is an exact result; it is not an approximation.

Function $\mathcal{M}\{\mathbf{a}, \mathbf{b}, \mathbf{s}, \mathbf{q} \mid \mathbf{z}\}$ plays the role of the function $g(y)$. We can now apply our Block Component Search Algorithm to the maximization of objective function $\mathcal{M}\{\mathbf{a}, \mathbf{b}, \mathbf{s}, \mathbf{q} \mid \mathbf{z}\}$. As in our earlier section, we begin by partitioning generic variable y into three components,

$$y_1 = \text{elements of } \mathbf{q}$$
$$y_2 = \text{elements of } \mathbf{a}, \mathbf{b}$$
$$y_3 = \text{elements of } \mathbf{s}.$$

By this decomposition, y_1 is associated with the event sequence, y_2 is associated with the wavelet's parameters, and y_3 is associated with the statistical parameters. Figure 4-7 summarizes a three step Block Component Search Algorithm. Its first step guarantees

$$\mathcal{M}\{a_i, b_i, s_i, q_i \mid z\} \leq \mathcal{M}\{a_i, b_i, s_i, q_{i+1} \mid z\} \; ;$$

its second step guarantees

$$\mathcal{M}\{a_i, b_i, s_i, q_{i+1} \mid z\} \leq M\{a_{i+1}, b_{i+1}, s_i, q_{i+1} \mid z\} \; ;$$

and, its third step guarantees

$$\mathcal{M}\{a_{i+1}, b_{i+1}, s_i, q_{i+1} \mid z\} \leq M\{a_{i+1}, b_{i+1}, s_{i+1}, q_{i+1} \mid z\} \; .$$

Combining these expressions, we see that

$$\begin{aligned}
\mathcal{M}\{a_i, b_i, s_i, q_i \mid z\} &\leq \mathcal{M}\{a_i, b_i, s_i, q_{i+1} \mid z\} \\
&\leq \mathcal{M}\{a_{i+1}, b_{i+1}, s_i, q_{i+1} \mid z\} \\
&\leq \mathcal{M}\{a_{i+1}, b_{i+1}, s_{i+1}, q_{i+1} \mid z\},
\end{aligned}$$

which demonstrates that the Block Component Search Algorithm is indeed increasing the function $\mathcal{M}\{\ \}$ from one iteration to the next.

The three generic variables y_1, y_2, and y_3 can be permuted to arrive at 6 different Block Component Search Algorithms. Again, there is no reason to believe that any one of these algorithms is better than any other. For that reason we shall continue to work with the Figure 4-7 algorithm.

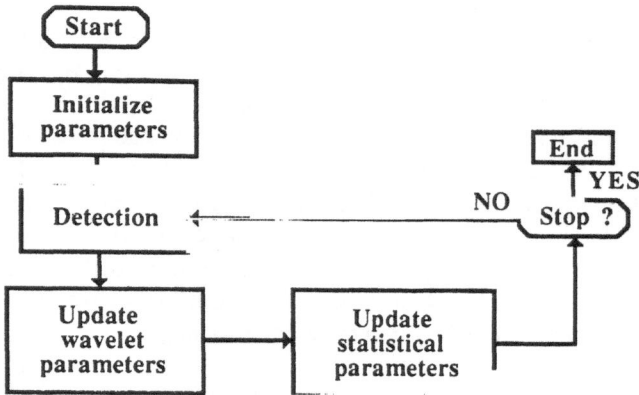

Figure 4-7. Three step Block Component Search Algorithm based on $\mathcal{M}$. Sub-iterations may occur within each major block.

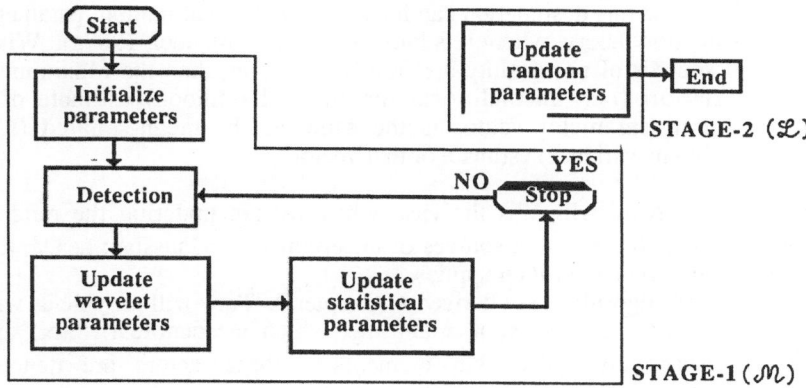

Figure 4-8. Two-stage likelihood optimization algorithm.

When we combine the three step Block Component Search Algorithm with the Separation Principle, we obtain the two-stage likelihood optimization algorithm that is depicted in Figure 4-8.

We turn next to a discussion of some of the methods that can be used to carry out the steps in a Block Component Search Algorithm.

4.6 Update Random Parameters

When we maximize the loglikelihood function $\mathcal{L}\{a, b, s, q, r, u_B \mid z\}$ using a Block Component Method, then r and u_B are computed using Equations (4-2) and (4-3) in which maximum-likelihood quantities are replaced by those obtained at the ith iteration, i.e.,

$$r_i = A(a_i, b_i, s_i, q_i)z \tag{4-4}$$

and

$$u_{B,i} = B(a_i, b_i, s_i, q_i)z \tag{4-5}$$

When we maximize the function $\mathcal{M}\{a, b, s, q \mid z\}$, then, according to the Separation Principle, we calculate r^{ML} and $u_B{}^{ML}$, from Equations (4-2) and (4-3), after $\mathcal{M}$ is maximized.

These algorithms for updating $\mathbf{r}$ and $\mathbf{u_B}$ are known as *Minimum-Variance Deconvolution* (MVD) algorithms (Mendel, 1981). The reason for this designation can be traced to classical random parameter estimation theory, where it is known that (e.g., Mendel, 1987a): When all sources of uncertainty are jointly Gaussian, then the Maximum a Posteriori (i.e., unconditional maximum likelihood) estimate of a random parameter vector is the same as the mean-squared (i.e., minimum-variance) estimate of that vector.

When $\mathbf{q}_i$ is given, which is the case when we are updating the random parameters, then all remaining sources of uncertainty are Gaussian; hence, this fact from estimation theory then applies.

Specific MVD algorithms are derived in Chapter 7. They will provide us with minimum-variance estimates of $\mathbf{r}$ and $\mathbf{u_B}$, which we denote $\mathbf{r}^{MV}(N)$ and $\mathbf{u_B}^{MV}(N)$, respectively. The kth elements of these vectors are denoted $\mathbf{r}^{MV}(k \mid N)$ and $\mathbf{u_B}^{MV}(k \mid N)$, in which the conditioning denotes the fact that all N measurements, z(1), z(2), ..., z(N), are used to obtain these estimates at each value of k (k = 1, 2, ... , N).

4.7 Binary Detection

Each element of the event vector $\mathbf{q}$ is a binary variable. When q(k) = 1 there is a seismic event at time point k, whereas, when q(k) = 0 there is no seismic event at time point k. At each time point we must make one of two decisions, there is or there is not an event. These decisions must be made in such a way that our loglikelihood function $\mathcal{L}\{\mathbf{a}, \mathbf{b}, \mathbf{s}, \mathbf{q}, \mathbf{r}, \mathbf{u_B} \mid \mathbf{z}\}$ [or function $\mathcal{M}\{\mathbf{a}, \mathbf{b}, \mathbf{s}, \mathbf{q} \mid \mathbf{z}\}$] increases.

The problem of having to make a choice between two decisions is called a *Binary Detection* problem. History does not record who first formulated such a problem precisely; but, perhaps it was Shakespeare's Hamlet when he said, "To be or not to be, that is the question."

Many strategies for binary detection are possible. Some process all the data just one time and lead to a decision rule about whether q(k) = 1 or 0 at all values of k = 1, 2, ... , N. These are *one-shot detectors*. Others process all of the data in iterative fashions. At each iteration there is a decision rule, often the same one, about whether q(k) = 1 or 0. These are *recursive detectors*. Recursive detectors are usually not self-starting, i.e., they need an initial set of values for q(k), k = 1, 2, ... , N. We use a subscript zero to denote the initial q(k) sequence, or the initial $\mathbf{q}$ vector , e.g., $\mathbf{q}_0$. Often, a one-shot detector provides us with just such an initial sequence. There are also other detectors that do not process all N measurements at once. Instead, they process one measurement at a time, beginning with z(1) and concluding with z(N). These are *sequential detectors* that usually find great application in real-time data processing situations. We shall focus our attention on non real-time detectors, because of our interest in the seismic deconvolution problem.

Regardless of the nature of the detector, we can view it as a three step procedure. First, we process the measurements z, usually in some nonlinear manner. We then create a *decision function*; it will be a function of the processed measurements. Then, we specify a *decision strategy*, one that permits us to decide whether $q(k) = 1$ or 0 at $k = 1, 2, \dots , N$. Let $D(z; k)$ denote the decision function. Note that at each value of time all N measurements are being used. Let $S_k[D(z; k)]$ denote the decision strategy . The subscript k on S lets us know that a decision must be made at each value of k.

Now for some specific detectors.

4.7.1 Threshold Detector

A threshold detector is a one shot detector. It could be used at any step in our Block Component Search Algorithm, but we advocate its use only the first time the detection step is called for. It initializes the q parameters. To run it, we must be given a_0, b_0, and s_0. A nonlinear function of the data is created by first processing all N measurements linearly, using a Minimum-Variance Deconvolution filter, to obtain $u^{MV}(k \mid N)$, $k = 1, 2, \dots , N$, and then squaring $u^{MV}(k \mid N)$. Then, we compute a time-varying threshold function $t(k)$, which depends in a rather complicated way on the error-variance between $u^{MV}(k \mid N)$ and $u(k)$. Our decision function is the difference between $[u^{MV}(k \mid N)]^2$ and $t(k)$, i.e.,

$$D(z; k) = [u^{MV}(k \mid N)]^2 - t(k). \qquad (4\text{-}6)$$

The threshold detector decision strategy is: at each value of k,

$$\text{If } [u^{MV}(k \mid N)]^2 - t(k) > 0 \quad \text{decide } q(k) = 1; \text{ or}$$
$$\text{If } [u^{MV}(k \mid N)]^2 - t(k) < 0 \quad \text{decide } q(k) = 0. \qquad (4\text{-}7)$$

If $[u^{MV}(k \mid N)]^2 - t(k) = 0$, we can decide $q(k) = 1$ or 0. The resulting event sequence is denoted q_{TD}. A derivation of this detector is given in Chapter 7.

An example of the threshold detector decision strategy is depicted in Figure 4-9. In Figure 4-9a we show both $[u^{MV}(k \mid N)]^2$ and the threshold function $t(k)$. Whenever $[u^{MV}(k \mid N)]^2$ exceeds the threshold, we decide that $q(k)$ equals unity. At all other points, we decide that $q(k)$ equals zero. These time points can be seen in Figure 4-9b.

The threshold function $t(k)$ can be adjusted in either an upward or downward manner by increasing or decreasing, respectively, the variance, v_B, of the backscatter, $u_B(k)$. This is shown in Figure 4-10 for the same data depicted in Figure 4-9a.

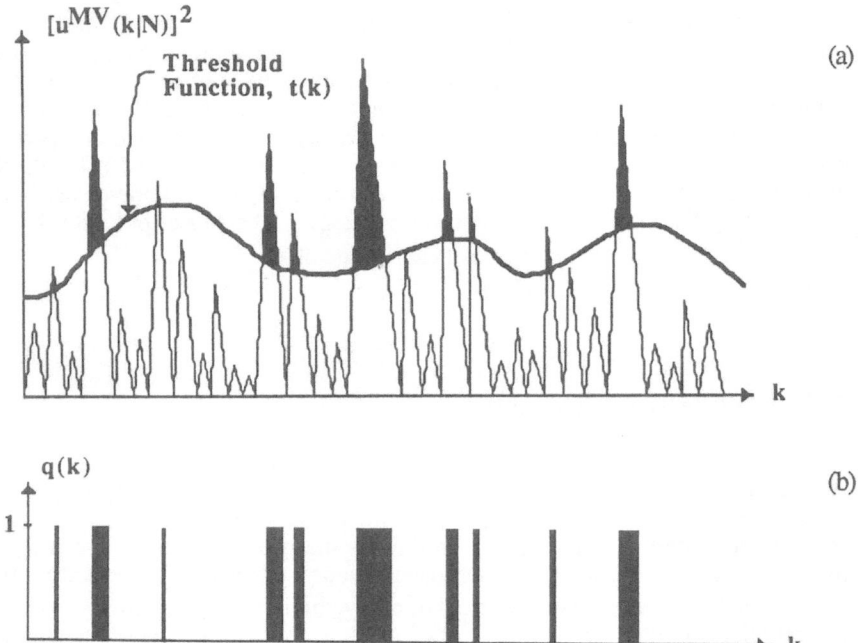

Figure 4-9. (a) Threshold detector decision strategy. (b) Times at which q(k) = 1 are in the black zones.

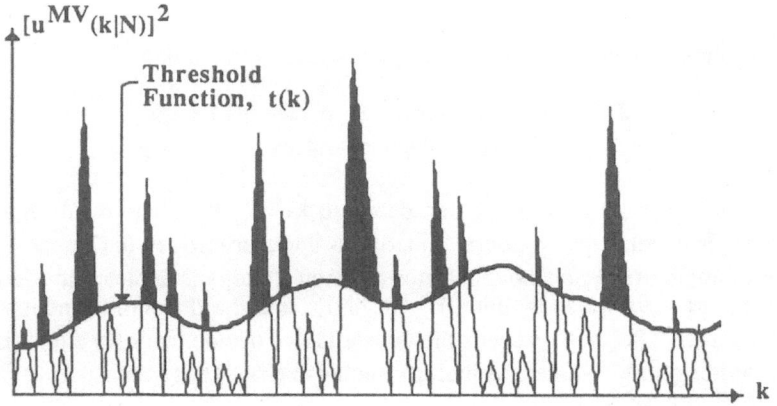

Figure 4-10.   By lowering the backscatter variance v_B the threshold function lowers (in a nonuniform manner), so that there are more time points at which q(k) = 1.

4.7.2 Single Most-Likely Replacement Detector

A single most-likely replacement detector (SMLR detector) is recursive (Kormylo, 1979, Kormylo and Mendel, 1982, or Mendel, 1983). It starts where the threshold detector leaves off. In order to understand the SMLR detector, consider the following experiment. We are given a_0, b_0, s_0, and a reference q, which we denote q_r. The first choice for q_r is the value of q obtained from the threshold detector, q_{TD}, or from other a priori knowledge. Recall that q_r is a vector with N elements. We now create N test vectors, each of which differs from q_r in exactly one location. These test vectors are denoted

$$q_{t,1}, q_{t,2}, \dots, q_{t,N}.$$

Figure 4-11 depicts test and reference event sequences for the simple case when N = 5 and q_r contains 3 nonzero elements.

Notice that the k th test vector, $q_{t,k}$, differs from q_r only at the k th time point.   By evaluating a likelihood ratio between $q_{t,k}$ and q_r, we obtain a decision function D(z;k). It processes the measurements in z in a very nonlinear manner. An example of this *likelihood-ratio decision function* is depicted in Figure 4-12. The SMLR detector decision strategy is:

> Find the value of k at which the likelihood-ratio decision function is a *maximum*. Call this time point k'. This will be the single time point at which a change is made in our reference sequence. The winning test sequence is $q_{t,k'}$.

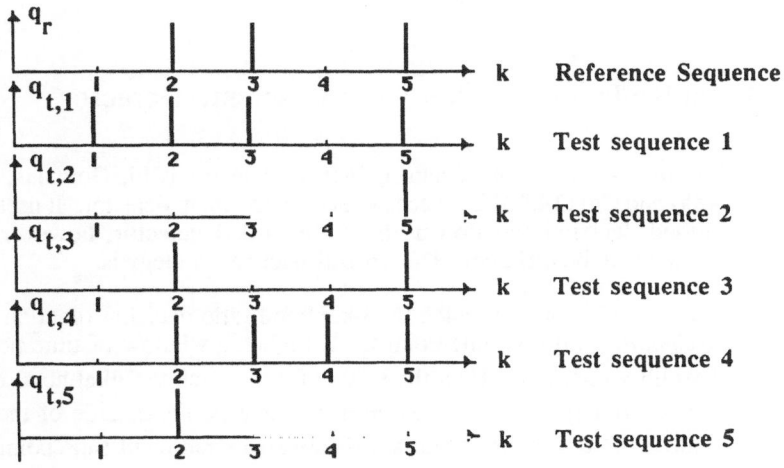

Figure 4-11. Reference and test sequences for the SMLR detector.

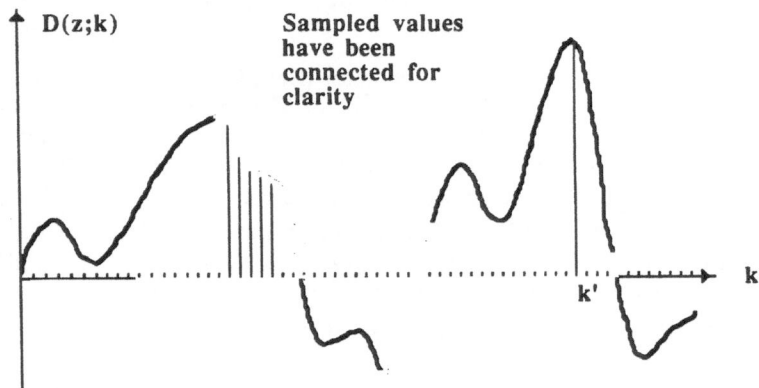

Figure 4-12. Likelihood-ratio decision function; k' is the winning value of k.

The SMLR detector is, as mentioned above, recursive. Once $q_{t,k'}$ has been found, this event vector replaces the original reference vector, and the procedure is repeated either for a fixed number of iterations, or until no test sequence can be found that is better than the most recent reference sequence. The latter occurs when the entire likelihood ratio decision function is negative.

Based on this description of the SMLR detector, and especially Figure 4-11, it appears that we will need N independent calculations in order to calculate the likelihood decision function D(z;k). Remarkably, this is not the case. In fact, *only one calculation is necessary*, and it involves just the reference sequence q_r. See Chapter 7 for details about D(z;k).

4.7.3 Multiple Most-Likely Replacement Detector

The multiple most-likely replacement (MMLR) detector [Chi, Goutsias, and Mendel (1985) and Chi (1987)] is an accelerated replacement detector. It uses the same likelihood decision function used by the SMLR detector; however, its decision strategy is quite different. The MMLR decision strategy is:

Find the value of k at which the likelihood ratio decision function is a *maximum*. Call this time point k_1. Establish a window of time points of width w_1 about k_1. Find the value of k at which the likelihood ratio decision function is a *maximum* at all time points outside of the w_1 window. Call this time point k_2. Establish a window of time points of width w_2 about k_2. Find the value of k at which the likelihood ratio decision function is a *maximum* at all time points outside of the w_1 and w_2 windows. Call this time point k_3. Continue this procedure until there are no time points that reside outside of the intersection of all windows. The winning sequence differs from the reference sequence at the multiple time points k_1, k_2, k_3, etc.

The resulting event vector is denoted q_w. This event vector replaces the original reference vector, and the procedure is repeated either for a fixed number of iterations, or until no sequence can be found that is better than the most recent reference sequence. The latter occurs when the entire likelihood ratio decision function is negative.

An example of the MMLR strategy is depicted in Figure 4-13. After k_1 is found, the window w_1 eliminates many points around k_1. The next winning point is at k_2. Window w_2 eliminates many more points. The next winning point at k_3 is located at the left-hand side of window w_1. Observe that window w_3 overlaps some of window w_1. This search procedure continues until no time points are left that are both uncovered by windows and have positive values for $D(z;k)$. There are 8 winning time points for this example.

The obvious advantage of the MMLR detector over the SMLR detector is its ability to make multiple changes in the reference sequence. This accelerates convergence of the detection block. It can happen, however, that the MMLR detector will only be able to make a change in one location of the reference sequence, in which case it functions as an SMLR detector. This behavior is data dependent.

How does one choose the windows w_1, w_2, etc.? There is no one best way. A *fixed window strategy* is to choose all of the windows of equal width. Unfortunately, no theory exists that demonstrates that this is a proper choice. An *adaptive window strategy* is to choose all of the windows of different width, where the actual widths are data dependent. While this is a more complicated

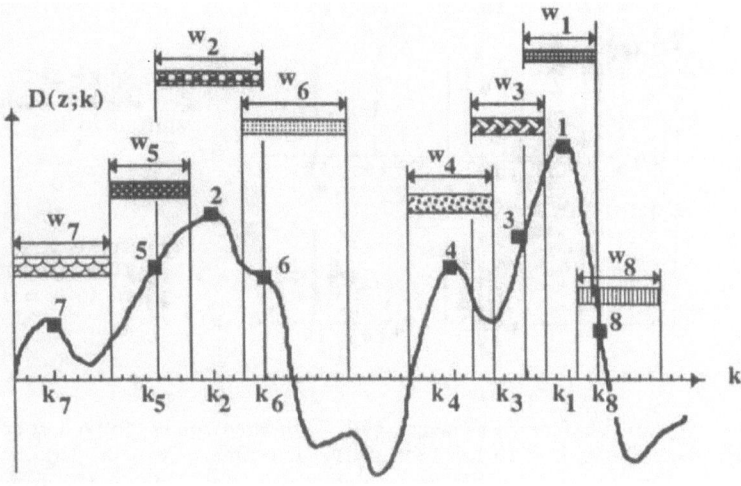

Figure 4-13. MMLR decision strategy.

strategy than the fixed window strategy, theory does exist that demonstrates that this is a proper choice (Chi, 1987).

4.7.4 Single Spike Shift Detector

Experience with the SMLR and MMLR detectors has shown that they sometimes lock onto local solutions that have clumps of spikes (i.e., at least two spikes that are one sample interval apart). This behavior is very data dependent, and occurs more frequently for low signal-to-noise ratios, and low bandwidth wavelets. An example that illustrates this behavior is depicted in Figure 4-14a. The single spike shift detector (SSS detector) has the capability to merge adjacent spikes, thereby providing even greater resolution than either the SMLR or MMLR detectors (Chi and Mendel, 1984a). In order to understand the SSS detector, consider the following experiment.

We are given a_0, b_0, s_0, and a reference sequence q, which we denote q_r. The first choice for q_r is the value of q obtained from any other detector, or from other a priori knowledge. Recall that q_r is a vector with N elements. Let us assume that q_r has exactly M unity elements. The exact locations of these

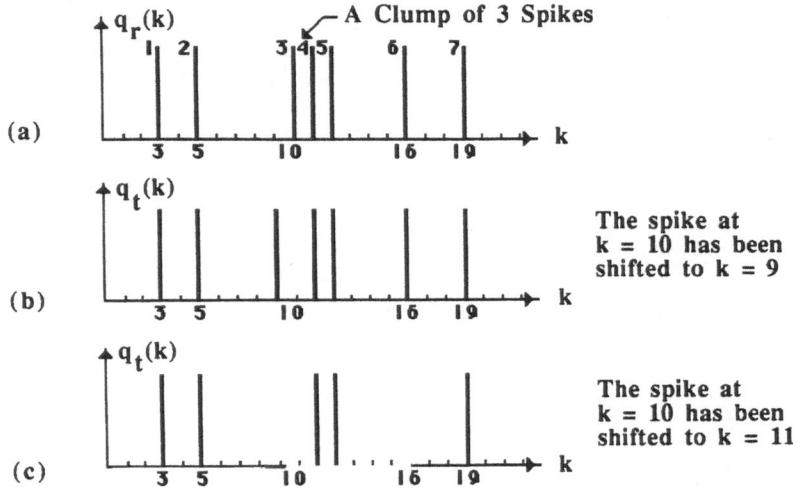

Figure 4-14. (a) A reference sequence with 7 ordered spikes; (b) A test sequence in which the spike at k = 10 has been shifted one time unit to the left, causing a change to occur in q_r at the two time points k = 9 and 10; (c) A test sequence in which the spike at k = 10 has been shifted one time unit to the right, causing a change to occur in q_r at the one time point k = 10.

elements is not important. We now create 2M test vectors, each of which differ from q_r in either one or two locations. Each of the M unity elements in q_r gives rise to two test vectors that are created in the following way. Let us focus our attention on a unity element located at time point $k = k_i$ (see Figure 4-14a, $k = 10$). The first test sequence that is associated with this point is obtained by *shifting* the unity spike one sampling point to the left to the time point $k = k_i$ - 1, whereas the second test sequence is obtained by *shifting* the unity spike one sampling point to the right to the time point $k = k_i + 1$. If the shifting produces a new unity spike, then the shifting operation will have caused a change to occur in q_r at *two* time points (see Figure 4-14b, $k = 10$). The test sequence will now be equal to zero at $k = k_i$ and unity at $k = k_i$ -1 or $k = k_i$ +1. Shifting does not have to produce a new spike. If a unity spike already exists at $k = k_i$ -1 or $k = k_i$ +1, then the shifted spike from time point $k = k_i$ merges with these already present spikes (see Figure 4-14c, $k = 10$). In this case the shifting operation causes a change to occur in q_r at only *one* time point.

By this procedure, we have constructed the test event vectors

$$q_{t,1}, q_{t,2}, \dots , q_{t,2M} .$$

As in the case of the SMLR detector, we now evaluate a likelihood ratio between each of these 2M test sequences and q_r, to obtain a decision function $D(z;i)$. It processes the measurements in a very nonlinear manner. For details about $D(z;i)$ see Chapter 7. An example of this likelihood-ratio decision function is depicted in Figure 4-15. The SSS detector decision strategy is:

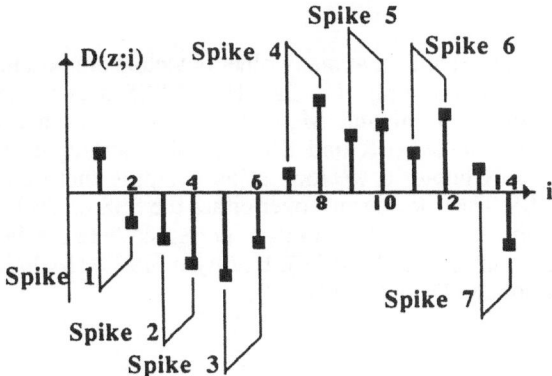

Figure 4-15. SSS likelihood-ratio decision function for the Figure 4-14a reference sequence. For each of the numbered spikes, the decision function is first computed for the test sequence in which the spike is shifted one unit to the left, and then it is computed for the test sequence in which the spike is shifted one unit to the right.

Find the value of i at which the likelihood-ratio decision function is a *maximum*. Call this time point i'. This maps into a single time point at which a change is initiated in our reference sequence. The winning test sequence is $q_{t,i'}$.

For the example in Figure 4-15, the winning test sequence occurs at i = 8. It is associated with shifting the fourth spike from k = 11 to k = 12; hence, the winning test sequence is $q_{t,8}$ and is the one that has unity spikes at k = 3, 5, 10, 12, 16, and 19.

The SSS detector is recursive, just as is the SMLR detector. Once $q_{t,i'}$ has been found, this event vector replaces the original reference vector, and the procedure is repeated either for a fixed number of iterations, or until no test sequence can be found that is better than the most recent reference sequence. The latter occurs when the entire likelihood-ratio decision function is negative.

There is one important difference between the SSS and SMLR likelihood-ratio decision functions. In each iteration of the SMLR detector we create a likelihood-ratio decision function that has exactly the same number of elements, namely N. In each iteration of the SSS detector, on the other hand, we create a likelihood-ratio decision function that may have a different number of elements, namely M. Recall that M is the number of unity elements in the reference sequence. Because the winning test sequence replaces the reference sequence, and the number of its unity elements may be different from the number in the reference sequence, M may vary from iteration to iteration.

4.7.5 Other Detectors

By now the reader must realize that many other detectors are possible. The most obvious one involves combining the SMLR and SSS decision strategies into an SSS-SMLR strategy (Chi and Mendel, 1984a). Another obvious one involves combining the MMLR and SSS decision strategies into an SSS-MMLR strategy. In Chapter 6 we look at an example that demonstrates the superiority of the SSS-SMLR detector over either the SSS or SMLR detectors.

Other detectors that are unrelated to those presented here can be found in the literature [e.g., Giannakis, et. al. (1989), Kormylo (1979), Mahalanabis, et. al. (1982), and Goussard and Demoment (1987)].

4.8 Update Wavelet Parameters

Loglikelihood function $\mathcal{L}\{a, b, s, q, r, u_B \mid z\}$ or objective function $\mathcal{M}\{a, b, s, q \mid z\}$ are very complicated and depend in very nonlinear ways on wavelet parameters a and b. The only known methods for maximizing these functions with respect to **a** and **b** are mathematical programming ones, that include: direct search, random search, search methods that use information about the first derivative of $\mathcal{L}\{\ \}$ or $\mathcal{M}\{\ \}$ (e.g., steepest descent, conjugate gradient, etc.), and search methods that use both first- and second-derivative information about $\mathcal{L}\{\ \}$ or $\mathcal{M}\{\ \}$. Newton-Raphson and Marquardt-Levenberg algorithms are examples of the latter.

Because the Marquardt-Levenberg algorithm has been widely applied to MLD (e.g., Kormylo, 1979 and Mendel, 1983), we describe it here. A brief derivation of this algorithm is given in Chapter 7. We assume that our objective is to maximize $\mathcal{M}\{\mathbf{a}, \mathbf{b}, \mathbf{s}, \mathbf{q} \mid \mathbf{z}\}$. Comparable discussions apply for the maximization of $\mathcal{L}\{\mathbf{a}, \mathbf{b}, \mathbf{s}, \mathbf{q}, \mathbf{r}, \mathbf{u}_B \mid \mathbf{z}\}$. Let $\mathbf{a}_i$ and $\mathbf{b}_i$ denote values of **a** and **b** obtained at the ith iteration of the Marquardt-Levenberg algorithm, and $\mathbf{a}_{i+1}$ and $\mathbf{b}_{i+1}$ denote values of **a** and **b** obtained at the i+1 st iteration. The Marquardt-Levenberg algorithms for **a** and **b** are (Marquardt, 1963):

$$\mathbf{a}_{i+1} = \mathbf{a}_i - (\mathbf{H}_{a,i} + \mathbf{D}_{a,i})^{-1}\mathbf{g}_{a,i} \qquad (4\text{-}8)$$

and

$$\mathbf{b}_{i+1} = \mathbf{b}_i - (\mathbf{H}_{b,i} + \mathbf{D}_{b,i})^{-1}\mathbf{g}_{b,i} . \qquad (4\text{-}9)$$

In these equations, $\mathbf{g}_{a,i}$ is the gradient of $\mathcal{M}\{\mathbf{a}, \mathbf{b}, \mathbf{s}, \mathbf{q} \mid \mathbf{z}\}$ with respect to **a** evaluated at $\mathbf{a} = \mathbf{a}_i$ and $\mathbf{b} = \mathbf{b}_i$, and $\mathbf{g}_{b,i}$ is the gradient of $\mathcal{M}\{\mathbf{a}, \mathbf{b}, \mathbf{s}, \mathbf{q} \mid \mathbf{z}\}$ with respect to **b** evaluated at $\mathbf{a} = \mathbf{a}_i$ and $\mathbf{b} = \mathbf{b}_i$, i.e.,

$$\mathbf{g}_{a,i} = \mathrm{col}[\partial \mathcal{M}\{\mathbf{a}, \mathbf{b}, \mathbf{s}, \mathbf{q} \mid \mathbf{z}\}/\partial a_j \mid \mathbf{a} = \mathbf{a}_i \text{ and } \mathbf{b} = \mathbf{b}_i; j = 1,...,n] \qquad (4\text{-}10)$$

and

$$\mathbf{g}_{b,i} = \mathrm{col}[\partial \mathcal{M}\{\mathbf{a}, \mathbf{b}, \mathbf{s}, \mathbf{q} \mid \mathbf{z}\}/\partial b_j \mid \mathbf{a} = \mathbf{a}_i \text{ and } \mathbf{b} = \mathbf{b}_i; j = 1,...,n]. \qquad (4\text{-}11)$$

Additionally, $\mathbf{H}_{a,i}$ is the Hessian matrix of $\mathcal{M}\{\mathbf{a}, \mathbf{b}, \mathbf{s}, \mathbf{q} \mid \mathbf{z}\}$ with respect to **a** (i.e., the matrix of second partial derivatives of $\mathcal{M}\{\mathbf{a}, \mathbf{b}, \mathbf{s}, \mathbf{q} \mid \mathbf{z}\}$ with respect to **a** evaluated at $\mathbf{a} = \mathbf{a}_i$ and $\mathbf{b} = \mathbf{b}_i$) and $\mathbf{H}_{b,i}$ is the Hessian matrix of $\mathcal{M}\{\mathbf{a}, \mathbf{b}, \mathbf{s}, \mathbf{q} \mid \mathbf{z}\}$ with respect to **b**, i.e.,

$$\mathbf{H}_{a,i} = \{\partial^2 \mathcal{M}\{\mathbf{a}, \mathbf{b}, \mathbf{s}, \mathbf{q} \mid \mathbf{z}\}/\partial a_j\, \partial a_m \mid \mathbf{a} = \mathbf{a}_i \text{ and } \mathbf{b} = \mathbf{b}_i;$$
$$j,m = 1,...,n\} \qquad (4\text{-}12)$$

and

$$\mathbf{H}_{b,i} = \{\partial^2 \mathcal{M}\{\mathbf{a}, \mathbf{b}, \mathbf{s}, \mathbf{q} \mid \mathbf{z}\}/\partial b_j\, \partial b_m \mid \mathbf{a} = \mathbf{a}_i \text{ and } \mathbf{b} = \mathbf{b}_i;$$
$$j,m = 1,...,n\} . \qquad (4\text{-}13)$$

Finally, $\mathbf{D}_{a,i}$ and $\mathbf{D}_{b,i}$ are diagonal stabilizing matrices; they are supposed to ensure the invertibility of matrices $(\mathbf{H}_{a,i} + \mathbf{D}_{a,i})$ and $(\mathbf{H}_{b,i} + \mathbf{D}_{b,i})$, respectively.   Note that matrices $(\mathbf{H}_{a,i} + \mathbf{D}_{a,i})$ and $(\mathbf{H}_{b,i} + \mathbf{D}_{b,i})$ must be positive definite for their inverses to exist (Stewart,1973).

In order to understand the Marquardt-Levenberg algorithm, we present an example.   Figure 4-16 shows equi-likelihood contours for a two parameter situation.   The gradient to the loglikelihood function is perpendicular to the tangent that is drawn at a specific point.   This specific point represents $a_{1,i}$ and $a_{2,i}$.   The gradient vector is in the direction of steepest ascent to the loglikelihood function.

When matrix $(\mathbf{H}_{a,i} + \mathbf{D}_{a,i})^{-1}$ multiplies $\mathbf{g}_{a,i}$, it not only changes the direction of $\mathbf{g}_{a,i}$, but it also changes the length of the gradient vector (the length of a vector equals the square root of the sum of the squares of its components).   An example of this is depicted in Figure 4-17.   The new vector, which is added to $\mathbf{a}_i$, is now hopefully pointing in a better direction than the gradient, so that convergence to the maximum of $\mathcal{M}\{\mathbf{a}, \mathbf{b}, \mathbf{s}, \mathbf{q} \mid \mathbf{z}\}$ will hopefully be accelerated.

Figure 4-18 depicts a flowchart for the Marquardt-Levenberg procedure. *Initialization* is accomplished by providing initial values for the wavelet parameters, say $\mathbf{a}_O$ and $\mathbf{b}_O$, as well as the wavelet's order n. Sometimes, we begin with a minimum phase initial wavelet that is obtained by a wavelet extraction procedure. This wavelet provides us with $\mathbf{a}_O$, $\mathbf{b}_O$ and n. Higher-order statistics can also be used to obtain $\mathbf{a}_O$, $\mathbf{b}_O$ and n (e.g., Tugnait, 1985, and Giannakis and Mendel, 1989). In fact, they should be used if we believe that the wavelet is nonminimum phase.   It is very important to obtain $\mathbf{a}_O$, $\mathbf{b}_O$ and n values that are associated with a correct phase initial wavelet. Such values will be 'closer' to the true values and will accelerate convergence of the Block Component Method. Sometimes the wavelet order, n, is treated as a variable, and MLD solutions are obtained for different values of n. See Mendel (1983, pp. 148-149) for discussions on how to do this. A good choice for n is crucial to the success of MLD.

If $(\mathbf{H}_{a,i} + \mathbf{D}_{a,i})$ and $(\mathbf{H}_{b,i} + \mathbf{D}_{b,i})$ are invertible, then $\mathbf{a}_i$ and $\mathbf{b}_i$ are *updated* using the just-described Marquardt-Levenberg algorithm. If, on the other hand, $(\mathbf{H}_{a,i} + \mathbf{D}_{a,i})$ and $(\mathbf{H}_{b,i} + \mathbf{D}_{b,i})$ are not invertible, then the following modified Marquardt-Levenberg algorithm is used:

$$\mathbf{a}_{i+1} = \mathbf{a}_i + \mathbf{D}_{a,i}^{-1}\mathbf{g}_{a,i} \qquad (4\text{-}14)$$

and

$$\mathbf{b}_{i+1} = \mathbf{b}_i + \mathbf{D}_{b,i}^{-1}\mathbf{g}_{b,i}. \qquad (4\text{-}15)$$

Because $\mathbf{D}_{a,i}$ and $\mathbf{D}_{b,i}$ are diagonal matrices, no problems occur in computing their inverses.

Having $\mathbf{a}_{i+1}$ and $\mathbf{b}_{i+1}$, we now *compute* $M\{\mathbf{a}_{i+1}, \mathbf{b}_{i+1}, \mathbf{s}, \mathbf{q} \mid \mathbf{z}\}$, and then *compare* it with $\mathcal{M}\{\mathbf{a}_i, \mathbf{b}_i, \mathbf{s}, \mathbf{q} \mid \mathbf{z}\}$. Recall, that our objective is to maximize $\mathcal{M}\{ \ \}$. If

$$\mathcal{M}\{\mathbf{a}_{i+1}, \mathbf{b}_{i+1}, \mathbf{s}, \mathbf{q} \mid \mathbf{z}\} > \mathcal{M}\{\mathbf{a}_i, \mathbf{b}_i, \mathbf{s}, \mathbf{q} \mid \mathbf{z}\},$$

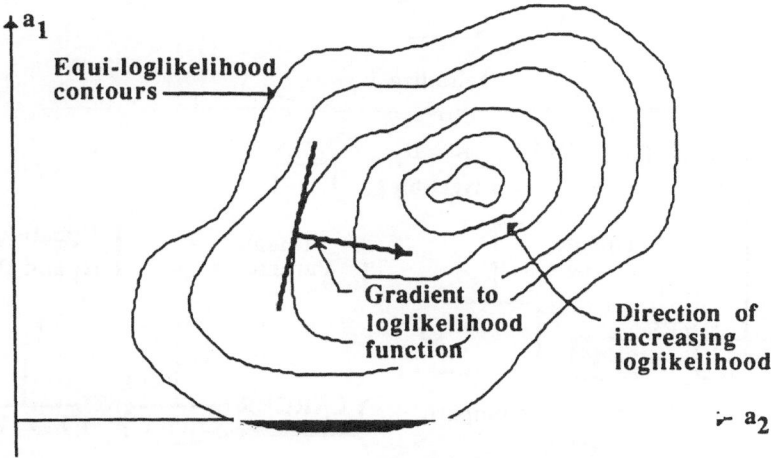

Figure 4-16. The gradient vector in relationship to the loglikelihood function.

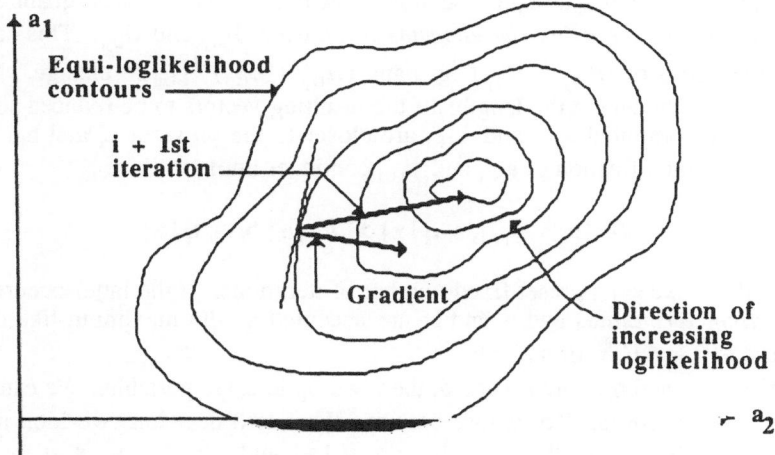

Figure 4-17. By altering the direction and length of the vector at the i+1 st iteration, the new vector is (hopefully) pointed in a better direction than the gradient vector.

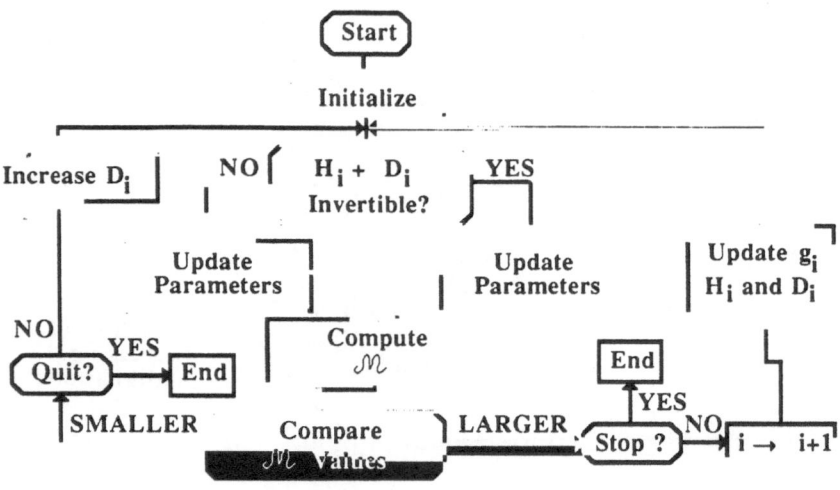

Figure 4-18. Flowchart for the Marquardt-Levenberg optimization procedure. Parameters are updated either by (NO path) (4-14) and (4-15) or (YES path) (4-8) and (4-9).

then we accept a_{i+1} and b_{i+1} and proceed. If, on the other hand,

$$\mathcal{M}\{a_{i+1}, b_{i+1}, s, q \mid z\} < \mathcal{M}\{a_i, b_i, s, q \mid z\},$$

we reject a_{i+1} and b_{i+1} and proceed to generate new choices for these quantities. We do this by *increasing the elements in matrices* $\mathbf{D}_{a,i}$ and $\mathbf{D}_{b,i}$. This causes the directions of $(\mathbf{H}_{a,i} + \mathbf{D}_{a,i})^{-1}g_{a,i}$ and $(\mathbf{H}_{b,i} + \mathbf{D}_{b,i})^{-1}g_{b,i}$ to change. More significantly, it causes the lengths of the resulting vectors to be reduced so that the newly computed a_{i+1} and b_{i+1} are closer to the previous a_i and b_i. This testing and modification of a_{i+1} and b_{i+1} continues until

$$\mathcal{M}\{a_{i+1}, b_{i+1}, s, q \mid z\} > \mathcal{M}\{a_i, b_i, s, q \mid z\},$$

or until we exceed a preset fixed number of iterations. If the latter occurs, the *algorithm terminates* and a_i and b_i are accepted as the maximum-likelihood values, a^{ML} and b^{ML}, of a and b.

When a_{i+1} and b_{i+1} are accepted, then two options are possible. We can *stop* or we can *continue*. To choose between these two decisions we look at the difference between $\mathcal{M}\{a_{i+1}, b_{i+1}, s, q \mid z\}$ and $\mathcal{M}\{a_i, b_i, s, q \mid z\}$. If $|\mathcal{M}\{a_{i+1}, b_{i+1}, s, q \mid z\} - \mathcal{M}\{a_i, b_i, s, q \mid z\}|$ is less than some preset small number, we stop the procedure and accept a_{i+1} and b_{i+1} as the maximum-likelihood values, a^{ML} and b^{ML}, of a and b. If $|\mathcal{M}\{a_{i+1}, b_{i+1}, s, q \mid z\} - \mathcal{M}\{a_i, b_i, s, q \mid z\}|$ is larger than this preset small number, we continue the

Marquardt-Levenberg procedure for at least one more iteration. This is done by *updating* the gradient vectors, Hessian matrices, and (possibly) the diagonal matrices, and returning to the place in our optimization procedure where we test the invertibility of the matrices $(\mathbf{H}_{a,i} + \mathbf{D}_{a,i})$ and $(\mathbf{H}_{b,i} + \mathbf{D}_{b,i})$.

For mathematical details on exactly how to compute the gradient vectors and Hessian matrices, see Chapter 7.

By means of the Marquardt-Levenberg optimization procedure, we are able to maneuver around in the **a-b** - parameter space, until we reach the maximum of $\mathcal{M}\{\ \}$. An example of this is depicted in Figure 4-19.

A word of caution is in order.   Convergence to the global maximum of $\mathcal{M}\{\mathbf{a, b, s, q} \mid \mathbf{z}\}$ is not guaranteed. We may lock onto a local maximum of $\mathcal{M}\{\mathbf{a, b, s, q} \mid \mathbf{z}\}$. Unless we restart the entire procedure from different values of $\mathbf{a}_0$ and $\mathbf{b}_0$, we will never know whether we have reached a local or global maximum. Our experiences with MLD indicate it doesn't matter much whether we have locked onto a local maximum instead of the global maximum. Of real importance is the convergence of the ML wavelet to the true wavelet, i.e., $\mathbf{W}^{ML} = \mathbf{W}(\mathbf{a}^{ML}, \mathbf{b}^{ML})$ to $\mathbf{W}_{TRUE}$, and this can occur for many different values of $\mathbf{a}^{ML}$ and $\mathbf{b}^{ML}$. This is discussed further in Chapter 5.

If the wavelet is known to be minimum phase, a wide variety of system identification techniques, such as recursive maximum-likelihood, extended least squares, and recursive instrumental variables, can be used to determine it [e.g., Kollias, et. al. (1985) and Kollias and Halkias (1985)]. If one knows ahead of time that the wavelet is one of a small number of *known* wavelets, then the maximum a posteriori approach described in Lainiotis, et. al. (1988b) is appropriate. All of these methods begin with functions other than L or M; hence, they do not fall within the scope of our unified approach.

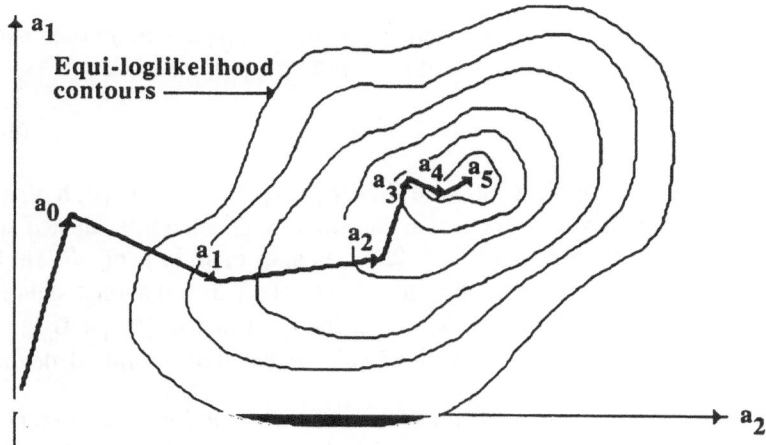

Figure 4-19. An example that illustrates the convergence of the updated parameters to the maximum of the loglikelihood function in a two-dimensional parameter subspace.

4.9   Update Statistical Parameters

Recall that there are four statistical parameters: v_r, the variance of the amplitude sequence; v_B, the variance of the backscatter sequence; v_n, the variance of the measurement noise sequence; and, λ, the average number of spikes in the event sequence. Regardless of whether we maximize $\mathcal{L}\{a, b, s, q, r, u_B \mid z\}$ or $\mathcal{M}\{a, b, s, q \mid z\}$, the three variances are always treated differently than the parameter λ.

Variances must always be positive; hence, updating variances to maximize either $\mathcal{M}\{\ \}$ or $\mathcal{L}\{\ \}$ is a *constrained optimization problem*. Such problems are inherently more difficult to solve than unconstrained optimization problems. We therefore avoid the direct approach, and choose instead to update the standard deviations rather than the variances. Standard deviations can be positive or negative; but, when they are squared to give variances the results are always positive. We are permitted to do this because maximum-likelihood quantitites enjoy the following *invariance property* (Mendel, 1987):

Functions of maximum-likelihood quantities
are themselves maximum-likelihood quantities.

Let p_1, p_2, and p_3 denote $v_r^{1/2}$, $v_B^{1/2}$, and $v_n^{1/2}$, respectively. We shall first obtain maximum-likelihood values of p_1, p_2, and p_3, denoted p_1^{ML}, p_2^{ML}, and p_3^{ML}. The invariance property of maximum likelihood lets us conclude that

$$v_r^{ML} = (p_1^{ML})^2, \quad v_B^{ML} = (p_2^{ML})^2, \text{ and } v_n^{ML} = (p_3^{ML})^2.$$

When $\mathcal{M}\{\ \}$ or $\mathcal{L}\{\ \}$ are maximized, p_1, p_2, and p_3 are all updated using a Marquardt-Levenberg algorithm. Let $\mathbf{p} = \text{col}(p_1, p_2, p_3)$; then,

$$\mathbf{p}_{i+1} = \mathbf{p}_i + (\mathbf{H}_{p,i} + \mathbf{D}_{p,i})^{-1} \mathbf{g}_{p,i}, \tag{4-16}$$

where: $\mathbf{g}_{p,i}$ is the gradient of $\mathcal{L}\{a, b, s, q, r, u_B \mid z\}$ or $\mathcal{M}\{a, b, s, q \mid z\}$ with respect to the three elements of $\mathbf{s}$, that we have denoted $\mathbf{p}$, evaluated at $\mathbf{p} = \mathbf{p}_i$; $\mathbf{H}_{p,i}$ is the Hessian matrix of $\mathcal{L}\{a, b, s, q, r, u_B \mid z\}$ or $\mathcal{M}\{a, b, s, q \mid z\}$ with respect to $\mathbf{p}$ evaluated at $\mathbf{p} = \mathbf{p}_i$; and $\mathbf{D}_{p,i}$ is a diagonal stabilizing matrix that is supposed to ensure the invertibility of matrix $(\mathbf{H}_{p,i} + \mathbf{D}_{p,i})$. The flowchart in Figure 4-18 is also applicable to the three standard deviation parameters in $\mathbf{p}$.

When both the wavelet and amplitude variance v_r are unknown, then v_r cannot be determined within a maximum-likelihood framework. Here we shall give a plausibility argument for this statement. See Chapter 7 for a mathematically correct argument (Goutsias and Mendel, 1986). The actual reasons for this are related to the specific formulas for the functions $\mathcal{M}$ or $\mathcal{L}$, and depend not only on the structure of the convolutional model but also on the apriori statistics. Maximizing $\mathcal{M}$ or $\mathcal{L}$ turns out to be an ill-posed problem with respect to parameter v_r. For different values of v_r we can obtain different solutions for all of the other parameters. See Chapter 5 for another function, $\mathcal{P}$, for which this is not true.

Recall the convolutional model in Eq. (2-13), i.e.,

$$z(k) = [r(k)q(k) + u_B(k)]*w(k) + n(k), \quad k = 1, 2, \dots, N. \quad (4-17)$$

This equation can also be written as

$$\begin{aligned} z(k) &= \{[r(k)/v_r^{1/2}]q(k) + u_B(k)/v_r^{1/2}\}*v_r^{1/2}w(k) + n(k) \\ &= [r'(k)q(k) + u_B'(k)]*w'(k) + n(k), \quad k = 1, 2, \dots, N, \quad (4-18) \end{aligned}$$

where

$$r'(k) = r(k)/v_r^{1/2} \quad (4-19)$$

$$u_B'(k) = u_B(k)/v_r^{1/2}, \quad (4-20)$$

and

$$w'(k) = v_r^{1/2}w(k). \quad (4-21)$$

From these equations we see that, *when $w(k)$, $r(k)$ and $u_B(k)$ are all unknown, we are in effect working with the scaled model in Eq.(4-18), in which we can at best determine the scaled wavelet $w'(k)$, the scaled amplitude sequence $r'(k)$, and the scaled backscatter sequence $u_B'(k)$.* Observe that the variance of the scaled amplitude sequence equals unity, i.e., $v_{r'} = 1$; hence, when all of these quantities are unknown we may just as well fix the variance of the amplitude sequence. It cannot be estimated.

Parameter λ is updated using the simple relaxation algorithm (Kormylo, 1979, and Mendel, 1983), that

$$\lambda_{i+1} = \text{(Number of unity elements in the last detected}$$
$$\text{event sequence, when the detector used } \lambda_i)/N. \quad (4-22)$$

In Chapter 7 we show that this algorithm is optimal in a maximum-likelihood sense.

Instead of updating all four of the statistical parameters together, we sometimes update λ after every stage of detection. This can be viewed, in any one of the previous Block Component Search Algorithms, as an adaptive detection loop. Figure 4-20 illustrates this for the Figure 4-8 two-stage likelihood optimization algorithm. The block that is labeled "Adaptive Detector" is one in which we iterate between detection and updating λ.

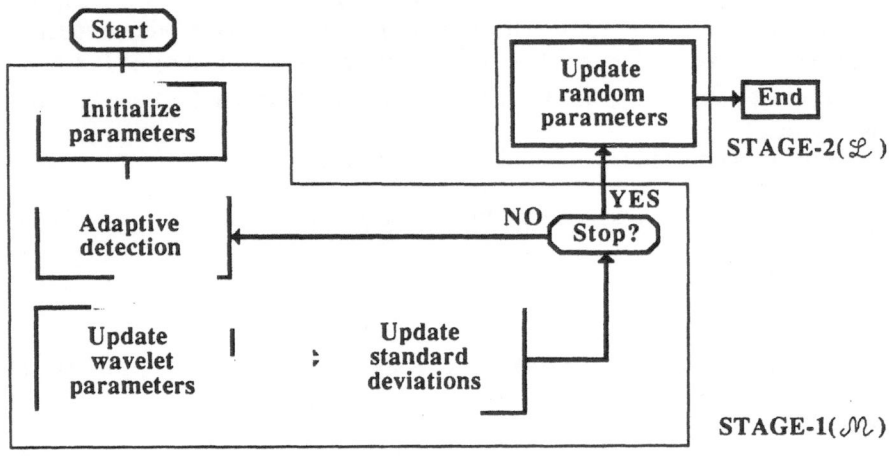

Figure 4-20.   Two-stage likelihood optimization algorithm with an adaptive detector.   Sub-iterations may occur within each major block.

From the brevity of this section, we see that updating the statistical parameters is quite straightforward.

4.10   Message For The Reader

You do not need to understand the derivations of the preceding major results in order to read Chapters 5 and 6. If you are planning to code MLD then you will need to extract specific formulas from Chapters 7 and 8; however, we strongly advise you to read Chapter 9 before doing such coding, because there are different ways to accomplish it, some of which are quite different than the formulas in Chapters 7 and 8 would have you believe. All of Chapter's 7 and 8's formulas are *batch formulas*. Often, it is computationally more expedient to implement *recursive* versions of them. These recursive versions are the subject of Chapter 9.

4.11   Summary

We have shown that our loglikelihood function, $\mathscr{L}$, or its related function, $\mathscr{M}$, can be maximized using Block Component Search Algorithms. The most general of these is depicted in Figure 4-20. Nonlinear signal processing is used in the blocks labeled "Adaptive Detection," "Update Wavelet Parameters" and "Update Statistical Parameters." Linear signal processing is used in the block labeled "Update Random Parameters." Using the structure of the Figure 4-20 algorithm, it is possible to code highly interactive software packages for accomplishing MLD.

Many examples that illustrate different aspects of MLD are given in Chapter 6.

Performance of MLD is limited by signal-to-noise ratio and the bandwidth of the seismic wavelet. Some aspects of performance are discussed in Chapter 5.

5
Properties and Performance

5.1 Introduction

In Chapter 4 we introduced three major types of signal processing: *minimum-variance deconvolution* (MVD), which is a form of linear signal processing; *detection*, which is a form of nonlinear signal processing; and, *optimization*, which is also a form of nonlinear signal processing. In this chapter we describe everything that is known to-date about the properties and performance of these three types of signal processing. We do this in order to gain a better appreciation and understanding of these techniques.

5.2 Minimum-Variance Deconvolution

In Chapter 1 we stated that "The design of a deconvolution operator requires a careful balancing of bandwidth and signal-to-noise ratio effects (see Figure 1-2)." In this section we shall examine the dependence of MVD on bandwidth and signal-to-noise ratio. We shall show that, under certain conditions, MVD leads to zero-phase waveshaped results, i.e., that the *resolution function* for MVD is a zero-phase function. Hence, not only does MVD perform linear deconvolution, but it simultaneously performs zero-phase waveshaping. The latter is precisely what interpretors prefer to view.

We shall also show that the MVD resolution function is sharpest for high signal-to-noise ratios and broad-band wavelets.

To begin, we need to state a formula for the MVD estimate of $u(k)$. The reader who is interested in mathematical details may find it useful at this point to jump ahead to the section in Chapter 7 entitled "Minimum-Variance Deconvolution." For those who are not interested in such details we pause to introduce some new notation, just as is done at the beginning of the Chapter 7 section on MVD.

Values of $u(k)$ that are a function of the data are known as *estimates* of $u(k)$. Minimum-variance estimates of $u(k)$ are denoted $u^{MV}(k \mid N)$. The conditioning denotes the fact that all N measurements $z(1)$, $z(2)$, ... , $z(N)$, are used to obtain estimates of each $u(k)$, $k = 1, 2, ... , N$. Here is the structure of $u^{MV}(k \mid N)$:

$$u^{MV}(k \mid N) = v_u(k)w_k'(WP_uW' + v_nI)^{-1}z, \qquad (5\text{-}1)$$

where $v_u(k)$ is the variance of $u(k)$, w_k is the k th column of wavelet matrix W given in (2-18), and P_u is the covariance matrix of the vector of all inputs $u = col(u(1), u(2), ... , u(N))$.

Observe that, at each value of k, $u^{MV}(k \mid N)$ depends on all N measurements, $z(1), z(2), ... , z(N)$. *This represents a noncausal dependence of the MVD filter on the measurements.* In Figure 5-1 the time axis is divided into two regions, in relation to the time point k. This time point is viewed as the "present." Of course, as k takes on different values the meaning of "present time" changes. If we glue ourselves to time point k, then the notions of past and future time points is the same, regardless of the specific values of k. *Past time points* are always to the left of time point k, whereas *future time points* are always to the right of k. It is the dependence of $u^{MV}(k \mid N)$ on future time points that introduces the noncausality.

To see this more clearly, we reexpress $u^{MV}(k \mid N)$, in Eq. (5-1), as

$$u^{MV}(k \mid N) = g(k,1)\, z(1) + g(k,2)z(2) + ... + g(k,k\text{-}1)z(k\text{-}1) + g(k,k)z(k)$$
$$+ g(k, k+1)z(k+1) + ... + g(k,k+N)z(N). \qquad (5\text{-}2)$$

The exact nature of the MVD filter coefficients is not important to this discussion. The first k-1 terms in Eq. (5-2) represent the effect of the past measurements on $u^{MV}(k \mid N)$. The k th term represents the effect of the present measurement on $u^{MV}(k \mid N)$. The last N-k terms represent the effect of the future measurements on $u^{MV}(k \mid N)$.

The MVD filter is, in general, a time-varying digital filter. In order to develop some insight into the expected performance of the MVD filter, we shall assume throughout the rest of this section that it is a constant coefficient filter. In this case, there is a totally different way to calculate the constant MVD filter coefficients from the way in which these coefficients would have to be calculated using Eq. (5-1) [i.e., by evaluating $v_u w_k'(WP_uW' + v_nI)^{-1}$]. It involves a different derivation of the MVD filter.

Consider the situation depicted in Figure 5-2. Measurements z(k) are processed by a two-sided digital filter. During the design of this filter its output, y(k), is compared to a desired output, d(k), and, the filter's coefficients are chosen to minimize the mean-squared error between d(k) and y(k). Naturally, in the design of a deconvolution filter, d(k) is chosen to be the input, u(k).

Filter output y(k) can be expressed as

$$y(k) = f(k)*z(k) = \sum_{i=-\infty}^{\infty} f(i)z(k\text{-}i) = ... + f(\text{-}2)z(k + 2) + f(\text{-}1)z(k + 1)$$

$$+ f(0)z(k) + f(1)z(k\text{-}1) + f(2)z(k - 2) + ... \qquad (5\text{-}3)$$

Comparing Eqs. (5-2) and (5-3), we see that (5-2) is a truncated version of (5-3). For our present approach, we are assuming a doubly-infinite two-sided filter, which means that the filter is described by the coefficients $f(0)$, $f(\pm1)$, $f(\pm2)$, ...

A derivation of these coefficients is given in Chapter 8, in the section entitled "MVD Filter Properties." There, it is shown that

$$F(\omega) = v_u W^*(\omega)/[v_u |W(\omega)|^2 + v_n], \qquad (5\text{-}4)$$

where $F(\omega)$ and $W(\omega)$ are the discrete-time Fourier transforms of $f(k)$ and $w(k)$, respectively, and $W^*(\omega)$ is the complex conjugate of $W(\omega)$.

The *resolution function*, $\rho(k)$, for this deconvolution filter is obtained by convolving the filter's input, $z(k)$, with $f(k)$, i.e.,

$$y(k) = f(k)^*z(k) = [f(k)^*w(k)]^*u(k) + f(k)^*n(k) = \rho(k)^*u(k) + f(k)^*n(k), \quad (5\text{-}5)$$

where

$$\rho(k) = f(k)^*w(k). \qquad (5\text{-}6)$$

Using the fact that $R(\omega) = F(\omega)W(\omega)$ and Eq. (5-4), we see that

$$R(\omega) = [|W(\omega)|^2 \, v_u/v_n]/[1 + |W(\omega)|^2 \, v_u/v_n] . \qquad (5\text{-}7)$$

Observe that *the resolution function is a zero-phase function.* Examples of such resolution functions are given in Figure 1-1. This means that *the noise-free portion of the deconvolution filter's output, namely* $\rho(k)^*u(k)$ *is a zero-phase waveshaped version of* $u(k)$. Hence, when the input is a stationary white sequence and the deconvolution filter is two sided doubly-infinite, then MVD zero-phase waveshapes as well as deconvolves (Chi and Mendel, 1984b).

Signal-to-noise ratio is directly proportional to the ratio v_u/v_n. Consequently, we see, from Eq. (5-7), that, as v_u/v_n approaches ∞, $R(\omega)$ approaches unity, so that $\rho(k)$ approaches the unit spike function; thus, for high signal-to-noise ratios the noise-free portion of the deconvolution filter's output approaches $u(k)$. This is the case of *perfect resolution.* Additionally, when $|W(\omega)|^2 v_u/v_n \gg 1$, $R(\omega)$ again approaches unity, so that once again perfect resolution is obtained. Broadband wavelets often satisfy this condition. The inequality $|W(\omega)|^2 v_u/v_n \gg 1$ clearly demonstrates the interrelationship between bandwidth and signal-to-noise ratio (see Figure 1-2).

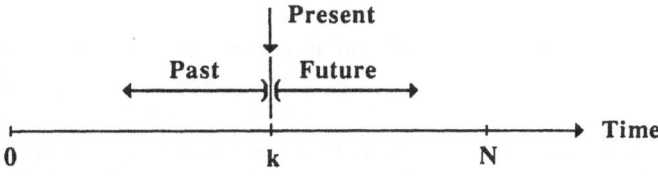

Figure 5-1. The notions of past, present and future values of time in relation to time point k.

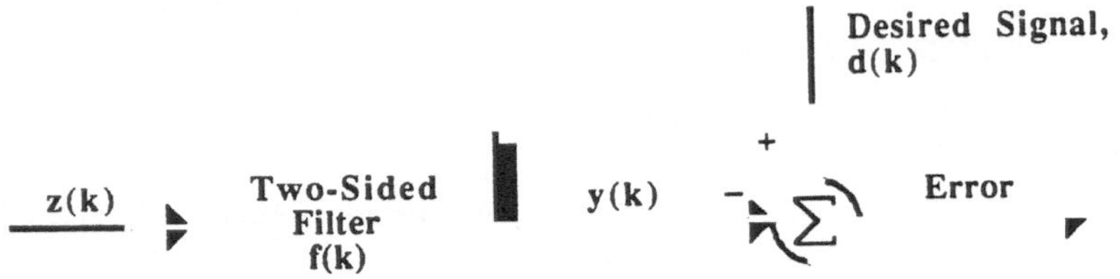

Figure 5-2. Design of two-sided MVD filter.

In general, $\rho(k)*u(k)$ will be a smeared out version of $u(k)$. The nature of the smearing is quite dependent on the bandwidth of $w(k)$ and signal-to-noise ratio. As mentioned in Chapter 1, "linear operations generally lead to more smeared out resolution functions than do nonlinear deconvolution operations." Minimum-Variance Deconvolution is no exception. Some specific examples of MVD resolution functions are given in Chapter 6.

In general, $\rho(k)*u(k)$ tends to undershoot $u(k)$. This is again due to bandwidth and signal-to-noise considerations. An example of this behavior is given in Chapter 6.

5.3   Detectors

All of the detectors that have been described in Chapter 4 are unconditional maximum-likelihood detectors (i.e., MAP detectors). Each of their decisions corresponds to exactly one of four possible situations:

1. The detector decides a reflector is present, and, indeed a reflector is present;

2. The detector decides no reflector is present, and, indeed a reflector is not present;

3. The detector decides a reflector is present, but in reality no reflector is present; and,

4. The detector decides no reflector is present, but in reality a reflector is present.

Clearly, situations 1 and 2 correspond to *correct decisions*, whereas situations 3 and 4 correspond to *errors*. Situation 3 is known as a *false alarm*, whereas situation 4 is known as a *miss*.

A different approach to designing a detector is to minimize the total probability of error, i.e., the combined probabilities of a false alarm and a miss. The following rather remarkable result is true: *The probability-of-error detection rule, obtained by minimizing the total probability of error, is identical to an unconditional maximum-likelihood detection rule* (Melsa and Cohn, 1978). Consequently, we can be assured that our MAP detectors will give rise to very few false alarms and misses.

5.3.1 Threshold Detector

In Chapter 4 we advocated initiating our recursive detectors with our *threshold detector*. Let $P_d(k)$ denote the probability of detection, i.e., the probability of being in situation 1, described above, and $P_f(k)$ denote the probability of false alarm, i.e., the probability of being in situation 3, described above. Formulas for $P_d(k)$ and $P_f(k)$ are derived in Chapter 8 for the threshold detector. These probabilities depend on our statistical parameters and MVD error variance in very complicated ways. In Chapter 8 we also show that

$$P_d(k) > P_f(k) \qquad\qquad (5\text{-}8)$$

While we cannot compare $P_d(k)$ and $P_f(k)$ analytically, other than by this inequality, they can be compared by means of simulations. In Example 6-6 we show that $P_d(k) >> P_f(k)$ and that $P_f(k)$ is very small, and, that these results are quite dependent on signal-to-noise ratio.

5.3.2 SMLR Detector

All of our recursive detectors guarantee an increase in likelihood. Their designs are predicated on this, and, they turn off when such an increase is no longer possible. It is important to remember though that they most likely will not achieve the global optimum of the loglikelihood function. They will lock onto a local maximum. Our experiences with these detectors is that such a local solution is usually good enough. This is not to say that if someone came up with a globally optimal MAP detection procedure we would not adopt it. If it was computationally efficient, we would jump at the opportunity to use it.

The SMLR detector that is based on $\mathcal{M}\{\mathbf{a}, \mathbf{b}, \mathbf{s}, \mathbf{q} \mid \mathbf{z}\}$ can run into some false alarm problems. This is demonstrated in Example 6-7. In Chapter 7, we state the SMLR likelihood-ratio decision function for choosing between sequences $q_{t,k}$ and q_r; it is

$$\begin{aligned}
\ln D(z;k) = {} & (w_k{}'P_z{}^{-1}z)^2/\{v_r{}^{-1}[q_{t,k}(k) - q_r(k)]^{-1} + w_k{}'P_z{}^{-1}w_k\} \\
& + 2[q_{t,k}(k) - q_r(k)]\ln[\lambda/(1 - \lambda)], \quad k = 1, 2, \ldots, N
\end{aligned} \qquad (5\text{-}9)$$

in which $\mathbf{P_z}$ is the covariance matrix of measurement vector $\mathbf{z}$. This formula was originally derived by Kormylo (1979) for the case when there is no backscatter. It remains structurally unchanged when backscatter is present. Backscatter only affects the details of the calculations of $\mathbf{w_k}'\mathbf{P_z}^{-1}\mathbf{z}$ and $\mathbf{w_k}'\mathbf{P_z}^{-1}\mathbf{w_k}$. In Chapter 7 we show that both of these quantities are computed from MVD filtering.

It is possible for the SMLR detector to fill in an event at every value of k, which, of course, is not desireable. Here is how this can happen. Suppose that $q_r(k) = 0$ and $q_{t,k}(k) = 1$, and $\lambda > 0.5$. Then $\mathsf{Ln}D(z;k) > 0$ for *all* possible measurement vectors, $\mathbf{z}$, because both terms on the right-hand side of Eq. (5-9) will *always* be positive. Put another way, $\mathsf{Ln}D(z;k) > 0$ independent of the data. The test sequence will become unity at *all* time points for which $q_r(k) = 0$ when $\lambda > 0.5$. Consequently the final test sequence becomes $\mathbf{q_t} = \mathrm{col}(1, 1, \dots , 1)$.

It has also been found that the SMLR likelihood ratio decision function in Eq. (5-9) produces many false alarms even when $\lambda < 0.5$ (Kwakernaak, 1980). These observations were made from cases when it was assumed that no backscatter is present, i.e., when $v_B = 0$. Kormylo and Mendel (1982, © 1982 IEEE) note that

> each detected event adds another non-trivial random variable to be estimated. In this respect the detection problem is similar to the model order-determination problem in system identification, where increasing the model order adds more unknown parameters to be estimated. It is well known that the likelihood criterion cannot solve the model order-determination problem; so, we should not be surprised that it cannot solve the simultaneous estimation and detection problem (Kwakernaak, 1980).

> In model-order determination, a common solution is to use a *modified likelihood criterion*, such as Akaike's information criterion (Akaike, 1974).

Before we follow a similar approach, we conjecture that *it is quite possible that the SMLR likelihood ratio decision function in Eq. (5-9) will not produce many false alarms if backscatter is included.* We have argued, in Chapter 4, that adding backscatter is equivalent to raising threshold levels (e.g., see Figures 4-9 and 4-10), which means that as backscatter is increased, fewer and fewer events will be detected. Usually, the false alarms associated with the SMLR likelihood ratio decision function in Eq. (5-9) are low amplitude events. They should disappear when backscatter is included. Of course our analysis of Eq. (5-9), which is given just below that equation, is true regardless of whether or not backscatter is present; but, remember it is for $\lambda > 0.5$. In seismic applications $\lambda << 0.5$. Example 6-7 confirms our conjecture.

5.4 A Modified Likelihood Function

Kormylo and Mendel (1982) have introduced a modified likelihood function that leads to detectors which have much fewer false alarms. Their modified likelihood function is obtained by treating the random parameter vectors $\mathbf{r}$ and $\mathbf{u}_B$ as "nuisance" parameters, and integrating them out of the problem (Edwards, 1972). We let $N\{\mathbf{a}, \mathbf{b}, \mathbf{s}, \mathbf{q} \mid \mathbf{z}\}$ denote the *modified likelihood function*; it is given by the expression

$$N\{\mathbf{a}, \mathbf{b}, \mathbf{s}, \mathbf{q} \mid \mathbf{z}\} = (2\pi)^{-N/2}|\Omega|^{-1/2} \exp(-\mathbf{z}'\Omega^{-1}\mathbf{z}/2)\lambda^{m(q)}(1 - \lambda)^{[N - m(q)]}$$

(5-10)

and *modified loglikelihood function* $\mathcal{N}\{\mathbf{a}, \mathbf{b}, \mathbf{s}, \mathbf{q} \mid \mathbf{z}\}$ is given as

$$\mathcal{N}\{\mathbf{a}, \mathbf{b}, \mathbf{s}, \mathbf{q} \mid \mathbf{z}\} = -\frac{N}{2}\ln 2\pi - \mathbf{z}'\Omega^{-1}\mathbf{z}/2 - \frac{1}{2}\ln|\Omega| + m(q)\ln(\lambda)$$
$$+ [N - m(q)]\ln(1 - \lambda)$$

(5-11)

A derivation of Eq. (5-10) is given in Chapter 8.

According to Edwards (1972), integrating nuisance parameters out of a problem is not always successful. Maximization of functions L and N (or $\mathcal{L}$ and $\mathcal{N}$) will give the same values for $\mathbf{a}$, $\mathbf{b}$, $\mathbf{s}$, and $\mathbf{q}$ only if we assume a special form of prior probability density functions for $\mathbf{r}$ and $\mathbf{u}_B$. Unfortunately, these special priors are unknown to us. Consequently, different results will be obtained when L or N are maximized (Goutsias and Mendel, 1986).

A recursive block optimization method for maximization of $\mathcal{N}$ is obtained by setting

$$x = \text{elements of } \mathbf{q}, \quad \text{and } y = \text{elements of } \mathbf{a}, \mathbf{b}, \mathbf{s}.$$

Keeping $\mathbf{a}$, $\mathbf{b}$, and $\mathbf{s}$ constant and maximizing $\mathcal{N}$ with respect to x can be accomplished by detection. Keeping $\mathbf{q}$ constant and maximizing with respect to y can be accomplished with a Marquardt-Levenberg algorithm. This two step approach is another Block Component Search Algorithm. Application of the Marquardt-Levenberg algorithm to $\mathcal{N}$ is more complicated than its application to $\mathcal{M}$, because of the term $\ln|\Omega|$ that appears in $\mathcal{N}$.

Vectors $\mathbf{r}$ and $\mathbf{u}_B$ are determined using the results of the Separation Principle, Eqs. (4-2) and (4-3), respectively.

5.5 An Objective Function

Wait one minute! Vectors $\mathbf{r}$ and $\mathbf{u}_B$ are determined by maximizing the loglikelihood function $\mathcal{L}\{\mathbf{a}, \mathbf{b}, \mathbf{s}, \mathbf{q}, \mathbf{r}, \mathbf{u}_B \mid \mathbf{z}\}$, whereas $\mathbf{a}$, $\mathbf{b}$, $\mathbf{s}$, and $\mathbf{q}$ are determined by maximizing the loglikelihood function $\mathcal{N}\{a, b, s, q \mid z\}$. We want a total solution for a, b, s, q, r, and u_B that maximizes a single function.

Interestingly enough, the solution procedure that we have just described actually maximizes one *objective function*, $\mathcal{P}\{a, b, s, q, r, u_B \mid z\}$, where, as is shown in Chapter 8,

$$\mathcal{P}\{a, b, s, q, r, u_B \mid z\} = N\ln 2\pi + \frac{N}{2}\ln(v_n v_r v_B) - \frac{1}{2}\ln|\Omega|$$
$$+ \mathcal{L}\{a, b, s, q, r, u_B \mid z\} . \qquad (5\text{-}12)$$

We refer to $\mathcal{P}$ as an objective function and not a loglikelihood function, because there is no natural likelihood function whose logarithm is $\mathcal{P}$.

The Separation Principle, which is associated with $\mathcal{L}$, applies as well to $\mathcal{P}$, and, the detectors obtained from $\mathcal{P}$ are the same as those obtained from $\mathcal{N}$. Additionally, the maximization of $\mathcal{P}$ or $\mathcal{N}$ with respect to amplitude variance v_r is not an ill-posed problem; but, we still cannot determine v_r when everything is unknown. In this case, regardless of our choice for v_r, we will always obtain the same solutions for all of the other unknown parameters. These facts and properties are all derived in Chapter 8.

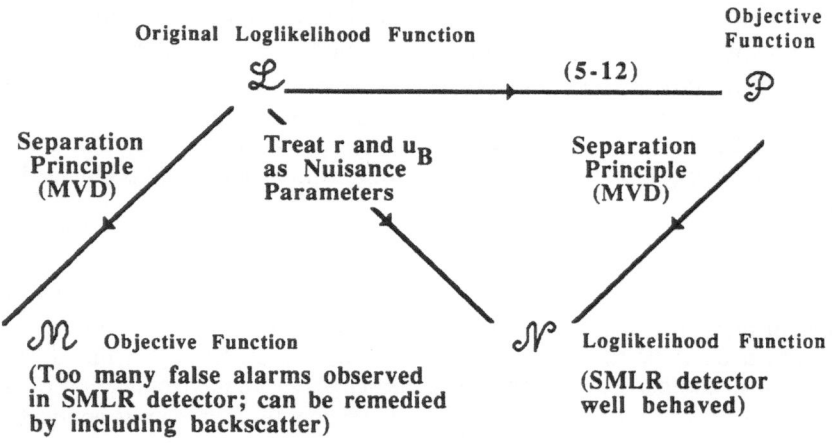

Figure 5-3. Interrelationships between loglikelihood functions $\mathcal{L}$ and $\mathcal{N}$, and, objective functions $\mathcal{M}$ and $\mathcal{P}$.

A different approach to the false-alarm problem of the SMLR detector is to develop a detector based on $\mathcal{L}$. Such a 'prediction-error detector' has been developed by Giannakis, Mendel and Zhao (1987). By their approach, $a, b, s, q,$ r, and u_B are all determined by focusing on the single loglikelihood function $\mathcal{L}$. Figure 5-3 summarizes the interrelationships between $\mathcal{L}, \mathcal{M}, \mathcal{N}$ and $\mathcal{P}$. As we have just mentioned, observe that the Separation Principle applies to both $\mathscr{L}$ and $\mathscr{P}$. Consequently, r and u_B are restored by MVD when working with either of these functions.

5.6 Marquardt-Levenberg Algorithm

The Marquardt-Levenberg algorithm is a modified Newton-Raphson algorithm. In Chapter 7, in the section entitled "Marquardt-Levenberg Algorithm," we state: "Let x_i denote the i th approximation to the maximum point [of a function $f(x)$]. To develop the Newton-Raphson algorithm we begin by performing a Taylor's series expansion of $f(x)$ about the point x_i.

The expansion of $f(x)$ up to the quadratic terms is

$$f(x) \simeq f(x_i) + [g_x(x_i)]'(x - x_i) - (x - x_i)'H_x(x_i)(x - x_i)/2, \qquad (5\text{-}13)$$

where $g_x(x_i)$ is the vector of first partial derivatives of f evaluated at x_i, and $H_x(x_i)$ is the matrix of second partial derivatives of f evaluated at x_i."

The Newton-Raphson algorithm has the following important property: *If f(x) is itself a quadratic function, we obtain the optimal solution x^* in a single step of the algorithm, because the Taylor's series expansion is exact.* This means that if we are close to the maximum and the function $f(x)$ is convex in the region of the maximum, so that a quadratic expansion exactly approximates the convex portion of $f(x)$, then convergence using the Newton-Raphson algorithm, and subsequently the Marquardt-Levenberg algorithm, is quite fast. It is "fast" in the sense that few iterations of the Marquardt-Levenberg algorithm will be required before we reach the (local) maximum. Each of these iterations may consume a significant amount of computation time because of the need to calculate first and second derivatives.

A cautionary word is in order about the Newton-Raphson method. It may diverge in some problems. It will converge to saddle points and to minima, as well as to maxima.

5.7 Convergence

When we use any one of the Chapter 4 Block Component Methods can we conclude anything about what our estimates will converge to? This is a very intriguing and difficult question to answer. We are dealing with random quantities, so that there are many different types of *stochastic convergence*, such as convergence in distribution, convergence almost everywheres, convergence in mean-square, and convergence in probability. Textbooks on probability theory (e.g., Papoulis, 1984) define these different modes of stochastic convergence and show how they are related. For example, convergence in mean-square implies convergence in probability, but the converse is not necessarily true.

The most popular and widely studied mode of stochastic convergence in estimation theory is convergence in probability. Estimate $c(k)$ of c converges in probability to c if

$$\lim_{k \to \infty} \Pr[\, | \, c(k) - c \, | \geqslant s] \text{ approaches } 0 \qquad (5\text{-}14)$$

where s is a small positive number. Such an estimator is known as a *consistent estimator*. Observe the limiting behavior, k approaches ∞, in the definition of convergence in probability. It means that convergence will occur as the number of measurements becomes infinite. This is a *large sample property* of an estimator. In practice this behavior will occur, if it occurs at all, for a "large" number of measurements. How "large" is problem and data dependent.

Convergence in probability is popular for at least two reasons. One is that it can be treated as an operator. For example, suppose that $x(k)$ and $y(k)$ are two random sequences, and that $x(k)$ converges in probability to a, whereas $y(k)$ converges in probability to b. Then the product sequence $x(k)y(k)$ converges in probability to ab, and the quotient sequence $x(k)/y(k)$ converges in probability to a/b, etc. This operator treatment of consistency often makes its study quite easy.

A second reason for the importance of consistency is the property that *consistency carries over*; i.e., any continuous function of a consistent estimator is itself a consistent estimator. For example, suppose $x(k)$ is a consistent estimator of x. Then $\sin[x(k)]$ is a consistent estimator of sin x and $\exp[x(k)]$ is a consistent estimator of exp x.

Recall that the convolutional model can be expressed as the vector measurement equation (2-15), i.e., as

$$\mathbf{z} = \mathbf{W}\mathbf{u} + \mathbf{n} \qquad (5\text{-}15)$$

where $\mathbf{u} = \text{col}(u(1), u(2), \dots , u(N))$. Regardless of what model we assume for the input $\mathbf{u}$, we may view it as a vector of random parameters. The k th element of $\mathbf{u}$, $u(k)$, is a random variable. Those readers who are familiar with the subject of parameter estimation, know that an equation like (5-15), which is linear in the vector of unknown random parameters $\mathbf{u}$, is the starting point for the derivations of many of the most famous parameter estimation methods, such as Least Squares, Mean-Squared, Maximum a Posteriori, and Best Linear Unbiased (e.g., Sorenson, 1980, Mendel, 1987). From this we conclude that *the design of a deconvolution signal processing filter is equivalent to the problem of estimating random parameters in a linear model*. This is why techniques from estimation theory, such as maximum likelihood, can be used to design deconvolution filters. It also means that everything that is known about convergence of different estimators of $\mathbf{u}$, from estimation theory, is applicable to our deconvolution problem. Unfortunately, there is a serious problem!

Although our MVD filter leads to mean-squared estimates of $\mathbf{u}$, $\mathbf{u}^{MV}(N)$ is not a consistent estimator of $\mathbf{u}$. Consistency is a large sample property of an estimator; however, as N increases, the dimension of $\mathbf{u}$ also increases, because $\mathbf{u}$ is N×1. Consequently, it is not possible to prove consistency of $\mathbf{u}^{MV}(N)$.

The fact that the design of a deconvolution filter is equivalent to estimating random parameters in a linear model tells us that, if someone designs a deconvolution filter and claims convergence of the estimated reflectivity to the true reflectivity in any stochastic mode that requires the number of measurements to approach infinity, we should treat such results with disbelief. The best we can hope to achieve in the design of a deconvolution filter is to provide its estimates with some desirable properties, such as minimum error-variance, maximum unconditional likelihood, minimum entropy, etc.

Let us now focus on the block, in Figure 4-20, that is labelled "Update Wavelet Parameters." Observe that the only unknowns for this block are the wavelet parameters contained in $\mathbf{a}$ and $\mathbf{b}$. No results are known about the convergence of a_i and b_i to their true values. While it would be terrific to know that a_i and b_i converge to their true values, this is not really necessary in deconvolution. In deconvolution we need to remove the effects of the wavelet from the data. An important question is "*Can we estimate the correct phase realization of wavelet w(k)?*" This question means we are interested in knowing whether the maximum-likelihood estimate of w(k), say $w^{ML}(k \mid N)$, converges in some sense to w(k) at all values of k. If it does, then we have captured the correct *shape* of the true wavelet and have therefore obtained a correct phase realization of w(k).

An answer to this question appears to be available only for loglikelihood function $\mathcal{N}\{\mathbf{a}, \mathbf{b}, \mathbf{s}, \mathbf{q} \mid \mathbf{z}\}$, under the rather strong assumption that $\mathbf{q}$ is known. One can view these results as *best case results* in the sense that they must deteriorate as estimates of $\mathbf{q}$ deteriorate. Of course, estimates of $\mathbf{q}$ are strongly dependent on signal-to-noise ratio and bandwidth.

The answer to this question was developed by Kormylo (1979) and Kormylo and Mendel (1983a), and is referred to by them as an *Identifiability Theorem*. We shall state this result as simply as possible. For all of the mathematical trappings, the reader should consult the two previous references, or Mendel (1983).

We assume that our wavelet is stable, finite-dimensional (which means that $\mathbf{a}$ and $\mathbf{b}$ are finite-dimensional), and time-invariant. Some other very technical assumptions, which are omitted here, are also needed. Choose $\mathbf{a}^{ML}$ and $\mathbf{b}^{ML}$ to maximize $\mathcal{N}\{\mathbf{a}, \mathbf{b}, \mathbf{s}, \mathbf{q} \mid \mathbf{z}\}$, under the assumption that $\mathbf{q}$ is known perfectly. Then

$$w^{ML}(k) = cw(k) \text{ for all } k = 0, 1, \dots \qquad (5\text{-}16)$$

where

$$c = \pm [v_r/v_r^{ML}]^{1/2} , \qquad (5\text{-}17)$$

if the true event sequence has the property that

$$\Phi_{qq}(\omega) \neq 0 \text{ for all values of } \omega, \tag{5-18}$$

in which

$$\Phi_{qq}(\omega) = \lim_{N \to \infty} \frac{1}{N} \sum_{k=1}^{N} \sum_{i=1}^{N} q(k)q(i) \, e^{j\omega(k-i)} . \tag{5-19}$$

This is an extremely interesting result. By testing the input event sequence, to see if it satisfies Eq. (5-18) we can determine if we will be able to reconstruct the correct phase realization of the wavelet. Because the condition in Eq. (5-18) is merely a *sufficient one* for identifiability of the wavelet, if it is not satisfied then no conclusions can be drawn. No conditions that are both necessary and sufficient are known at this time.

A sequence that satisfies Eq. (5-18) is called *persistently exciting*. Such a sequence is rich enough in frequencies so as to excite all of the modes of the wavelet. In essence, then, none of the wavelet will be invisible to the maximum-likelihood estimation procedure.

Observe, also, that Eq. (5-16) is a nonparametric result. It does not require an underlying parametric ARMA model for its proof.

Scale factor c represents an ambiguity that is present when the variance of the amplitude sequence, v_r, is unknown. Observe that, even if we know v_r (i.e., $v_r^{ML} = v_r$), there will still be a sign ambiguity in our estimate of $w(k)$, because $c = \pm 1$. This makes sense because we started with one equation in two unknowns, namely $w(k)$ and $u(k)$. If we are able to estimate the shape of $w(k)$ correctly, then its sign can either be absorbed correctly into it, or, it can be transferred into input $u(k)$. There is no way to determine which of these possibilities is the correct one, unless we are given more information. Sometimes we can resolve this ambiguity by looking at the data.

Quantity $\Phi_{qq}(\omega)$ looks like a *power spectrum*, but, strictly speaking, this is correct only when $q(k)$ is ergodic in autocorrelation [i.e., when ensemble averages of $q(i)q(i+j)$ equal temporal averages of this quantity (e.g., Papoulis, 1984)]. If $q(k)$ is ergodic in autocorrelation, then

$$\Phi_{qq}(\omega) = \sum_{l=-\infty}^{\infty} r_{qq}(l)e^{-j\omega l} , \tag{5-20}$$

where $r_{qq}(l)$ is the autocorrelation of $q(k)$. Observe that $\Phi_{qq}(\omega)$ is the discrete Fourier transform of $r_{qq}(l)$.

Now for the bottom line. When $q(k)$ is known to be a Bernoulli sequence, then it is ergodic in autocorrelation (Breiman, 1968). Additionally, as long as $\lambda \neq 0$ or $\lambda \neq 1$, then it is easy to show that $\Phi_{qq}(\omega) \neq 0$ for all ω. Under these conditions, which are easily met in practice, we conclude, from the Identifiability Theorem, that we will be able to identify wavelet $w(k)$ to within a scale factor of its true value.

Some recent and related identifiability results are given in Chu and Wang (1985).

A study into the convergence of the simple iterative update algorithm (4-22) is very intricate and is given in Kormylo (1979). The main result is that iterative update (4-22) converges monotonically to a local maximum of loglikelihood function $\mathscr{N}$ in N (here 'N' denotes the number of measurements, rather than $\exp\mathscr{N}$) or less steps for any initial value of λ_0 chosen within the closed interval from zero to unity. This result is only true under the assumption of *optimal* detection. When we use the SMLR detector, then alternating iterations of the SMLR detector and update (4-22) also converges in a finite number of steps. A simple proof of this fact is given in Chapter 8.

5.8 Entropy Interpretation

Recently, Giannakis and Mendel (1987) provided an entropic interpretation to MLD. They begin by observing, that: "From a statistical point of view, the output of a linear time invariant (LTI) system (in our case the seismogram) is a time averaging of the input (e.g., seismic reflectivity), the time averaging representing the system (e.g., the wavelet). In terms of the Central Limit Theorem, this averaging tends to give to the output a 'more Gaussian' nature than the one present in the original input. The task of seismic deconvolution is therefore to remove the averaging and produce the reflectivity, i.e., to move the seismogram towards 'less Gaussian' distributions. Equivalently, the task of seismic deconvolution is to maximize some measure of 'non-Gaussianity.'

"For a given variance of a random variable, the Gaussian distribution has maximum entropy, e.g., Papoulis (1984). Consequently, seismic deconvolution works toward a minimum entropy or maximum information direction, to produce the reflectivity ... .

"Because the measures of 'non-Gaussianity' and 'minimum entropy' are intuitive rather than rigorous, different authors [e.g., Wiggins (1978), Godfrey (1978), Claerbout (1978), Gray (1979), Ooe and Ulrych (1979), Claerbout (1979), Godfrey and Rocca (1981), and Deeming (1984)] have proposed different statistical entropy measures to model them. All these models and the resulting deconvolution techniques are known, under the name -- introduced by Wiggins (1978) -- as Minimum Entropy Deconvolution (MED). ...

"Most MED methods use an LTI (Linear Time-Invariant) deconvolution filter to produce the reflectivity; but they require nonlinear operations on the seismogram to optimize their entropy measure. ...

"The LTI deconvolution filter used in MED methods is obtained by maximizing an objective function, which leads to a highly nonlinear Toeplitz equation. The iterative maximization of this objective function, which corresponds to the minimization of a specific entropy measure, consists of successively estimating the reflectivity and inverse filtering. Iterations continue until convergence occurs....

"... all present MED methods do not explicitly include the presence of noise in the convolutional model. They consider the suppression of small level amplitude events the same as suppression of noise level events."

Giannakis and Mendel show how entropy concepts are embodied in MLD through the modeling parameter λ. They provide an entropy interpretation to the Figure 4-20 blocks labelled "Adaptive Detection" and "Update Random Parameters." In so doing they demonstrate why MLD asymptotically minimizes the entropy of the reflectivity sequence. In addition to this asymptotic behavior, they show that maximizing the loglikelihood ratio associated with detection corresponds to maximizing a cross-entropy. This puts MLD in the same framework as existing "adaptive" MED methods (Walden, 1985). They also show that the relaxation algorithm used for updating λ, given in Eq. (4-22) also minimizes a cross-entropy. Finally, because MLD includes noise in the convolutional model and can include a priori information about event locations, it is robust and can avoid geophysically unreasonable solutions [Deeming (1984) observed that MED methods can lead to a "geophysically unreasonable" optimum solution for the reflectivity; i.e., we can end up with a reflectivity whose amplitudes are all zero or unity].

5.9 Summary

Much is known about the performance of the different blocks in the Figure 4-20 Block Component Method. Invariably, the factors that limit the performance of each block are signal-to-noise ratio and bandwidth.

Minimum-variance deconvolution zero-phase waveshapes as well as deconvolves. In fact, one way in which MLD can be used is to go through its entire procedure just to get the information necessary to run the MVD filter. That information is wavelet parameters **a** and **b**, and statistical parameters **s**. Examples which illustrates this approach are given in Chapter 6 (Examples 6-1 and 6-2).

The recursive nature of our detectors guarantees that at each iteration the loglikelihood function will increase. No performance limits are known for the SMLR detector.

The Marquardt-Levenberg algorithm will converge in a small number of iterations when we are in the vicinity of a local extremum. This is because it is a variant of the Newton-Raphson method which is quadratically convergent.

Extensive simulations indicate that the Figure 4-20 Block Component Method leads to a wavelet estimate that is of the correct shape. This means that we have converged to a correct phase realization of the wavelet. If our method only guaranteed convergence to a minimum phase wavelet then there would be no value to it, because many other wavelet extraction methods do just that. By exploiting the nonGaussian nature of the input to the convolutional model, we are able to extract nonminimum phase wavelets.

MLD is also known to be an adaptive MED method. Its advantage over more conventional MED methods is that it accounts for noise whereas the conventional methods do not.

Finally, we return to Figure 5-3, which summarizes the interrelationships between $\mathcal{L}$, $\mathcal{M}$, $\mathcal{N}$, and $\mathcal{P}$. For future reference, we list these four functions, each in two different ways:

$$
\begin{aligned}
L\{\mathbf{\dot{a}}, \mathbf{b}, \mathbf{s}, \mathbf{q}, \mathbf{r}, \mathbf{u}_B \mid \mathbf{z}\} = {} & (2\pi)^{-3N/2}(v_n v_r v_B)^{-N/2} \exp[-\mathbf{r'r}/2v_r \\
& - (\mathbf{z} - \mathbf{WQr} - \mathbf{Wu}_B)'(\mathbf{z} - \mathbf{WQr} - \mathbf{Wu}_B)/2v_n \\
& - \mathbf{u}_B'\mathbf{u}_B/2v_B]\lambda^{m(q)}(1 - \lambda)^{[N - m(q)]} \qquad (5\text{-}21)
\end{aligned}
$$

$$
\begin{aligned}
\mathcal{L}\{\mathbf{a}, \mathbf{b}, \mathbf{s}, \mathbf{q}, \mathbf{r}, \mathbf{u}_B \mid \mathbf{z}\} = {} & -\tfrac{3N}{2} \ln 2\pi - \tfrac{N}{2} \ln(v_n v_r v_B) - \mathbf{r'r}/2v_r \\
& - (\mathbf{z} - \mathbf{WQr} - \mathbf{Wu}_B)'(\mathbf{z} - \mathbf{WQr} - \mathbf{Wu}_B)/2v_n - \mathbf{u}_B'\mathbf{u}_B/2v_B \\
& + m(q)\ln(\lambda) + [N - m(q)]\ln(1 - \lambda), \qquad (5\text{-}22)
\end{aligned}
$$

$$
\begin{aligned}
M\{\mathbf{a}, \mathbf{b}, \mathbf{s}, \mathbf{q} \mid \mathbf{z}\} = {} & (2\pi)^{-3N/2}(v_n v_r v_B)^{-N/2} \exp(-\tfrac{1}{2}\mathbf{z}'\Omega^{-1}\mathbf{z}) \\
& \lambda^{m(q)}(1 - \lambda)^{[N - m(q)]} \qquad (5\text{-}23)
\end{aligned}
$$

$$
\begin{aligned}
\mathcal{M}\{\mathbf{a}, \mathbf{b}, \mathbf{s}, \mathbf{q} \mid \mathbf{z}\} = {} & -\tfrac{3N}{2} \ln 2\pi - \tfrac{N}{2} \ln(v_n v_r v_B) - \tfrac{1}{2}\mathbf{z}'\Omega^{-1}\mathbf{z} \\
& + m(q)\ln(\lambda) + [N - m(q)]\ln(1 - \lambda), \qquad (5\text{-}24)
\end{aligned}
$$

$$
\begin{aligned}
N\{\mathbf{a}, \mathbf{b}, \mathbf{s}, \mathbf{q} \mid \mathbf{z}\} = {} & (2\pi)^{-N/2} \mid \Omega \mid^{-1/2} \exp(-\tfrac{1}{2}\mathbf{z}'\Omega^{-1}\mathbf{z}) \\
& \lambda^{m(q)}(1 - \lambda)^{[N - m(q)]} \qquad (5\text{-}25)
\end{aligned}
$$

$$
\begin{aligned}
\mathcal{N}\{\mathbf{a}, \mathbf{b}, \mathbf{s}, \mathbf{q} \mid \mathbf{z}\} = {} & -\tfrac{N}{2} \ln 2\pi - \tfrac{1}{2} \mathbf{z}'\Omega^{-1}\mathbf{z} - \tfrac{1}{2} \ln \mid \Omega \mid + m(q)\ln(\lambda) \\
& + [N - m(q)]\ln(1 - \lambda), \qquad (5\text{-}26)
\end{aligned}
$$

$$
\begin{aligned}
P\{\mathbf{a}, \mathbf{b}, \mathbf{s}, \mathbf{q}, \mathbf{r}, \mathbf{u}_B \mid \mathbf{z}\} = {} & (2\pi)^{N}(v_n v_r v_B)^{N/2} \mid \Omega \mid^{-1/2} \\
& L\{\mathbf{a}, \mathbf{b}, \mathbf{s}, \mathbf{q}, \mathbf{r}, \mathbf{u}_B \mid \mathbf{z}\} \qquad (5\text{-}27)
\end{aligned}
$$

$$
\begin{aligned}
\mathcal{P}\{\mathbf{a}, \mathbf{b}, \mathbf{s}, \mathbf{q}, \mathbf{r}, \mathbf{u}_B \mid \mathbf{z}\} = {} & N\ln 2\pi + \tfrac{N}{2}\ln(v_n v_r v_B) - \tfrac{1}{2}\ln \mid \Omega \mid \\
& + \mathcal{L}\{\mathbf{a}, \mathbf{b}, \mathbf{s}, \mathbf{q}, \mathbf{r}, \mathbf{u}_B \mid \mathbf{z}\} \qquad (5\text{-}28)
\end{aligned}
$$

Interestingly enough, there is a very strong relationship between N and M (and, subsequently, $\mathcal{N}$ and $\mathcal{M}$). From Equations (5-23) and (5-25), we find that

$$
M\{\mathbf{a}, \mathbf{b}, \mathbf{s}, \mathbf{q} \mid \mathbf{z}\} = (2\pi)^{-N} \mid \Omega \mid^{1/2} (v_n v_r v_B)^{-N/2} N\{\mathbf{a}, \mathbf{b}, \mathbf{s}, \mathbf{q} \mid \mathbf{z}\} \qquad (5\text{-}29)
$$

hence,

$$\mathcal{M}\{a, b, s, q, | z\} = -N\ln 2\pi - \frac{N}{2}\ln(v_n v_r v_B)$$
$$+ \frac{1}{2}\ln|\Omega| + \mathcal{N}\{a, b, s, q, | z\} \qquad (5\text{-}30)$$

Comparing M to objective function P, we see that they are quite similar. Just as P is related to likelihood function L, M is related to likelihood function N. There appears to be a duality between loglikelihood functions $\mathcal{L}$ and $\mathcal{N}$, and objective functions $\mathcal{P}$ and $\mathcal{M}$. This is epitomized by the identity [obtained by multiplying Eq. (5-27) by Eq. (5-29)] that MP = LN. Using the relationship between $\mathcal{M}$ and $\mathcal{N}$, we modify Fig. 5-3 to the final structure depicted in Fig. 5-4.

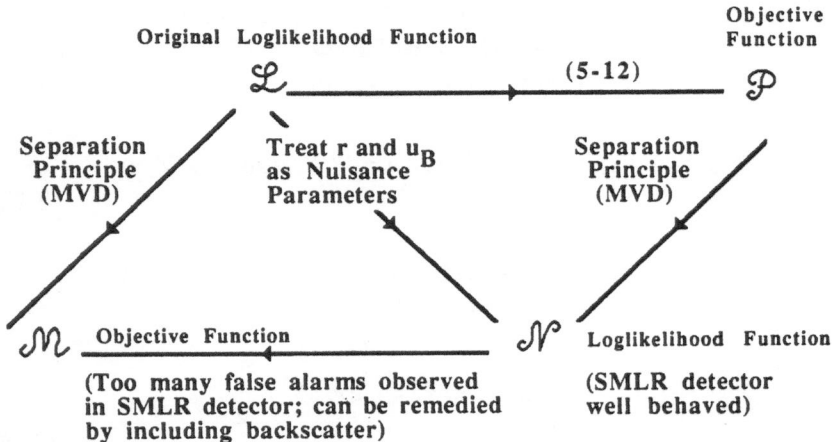

Figure 5-4. Total interrelationships between loglikelihood functions $\mathcal{L}$ and $\mathcal{N}$, and objective functions $\mathcal{P}$ and $\mathcal{M}$.

6
Examples

6.1 Introduction

This chapter is filled with examples that hopefully illuminate much of the material that has been described in Chapters 1 through 5, especially the material in Chapters 4 and 5. Real and synthetic data cases are presented, because much can be learned about MLD from both types of data. Rather than collect all of the real data examples in one section, at the end of the chapter, as is customarily done in journal articles, we shall weave them in with the synthetic data examples. In fact, we begin with some real data examples that illustrate the high resolution processing capabilities of MLD.

6.2 Some Real Data Examples

Three examples are given in this section. They are taken from Chi, et al. (1984) and were performed at Veritas Seismic, Ltd. by Dan Hampson and his colleagues.

In applying MLD to real data, it is essential to devise strategies that reduce the computational effort. In particular, estimates of the statistical parameters or the detected sequence may be assumed constant for particular data sets, depending upon whether the deconvolution is performed before or after stack.

In processing Common Depth Point stacked data, the following assumptions can often be made (Chi, et al., 1984):

1. signal-to-noise ratio is roughly constant across the section within a given window;

2. the rate of occurrence of geologic boundaries (λ) is constant; and,

3. the wavelet variations, from trace-to-trace, are small, though significant for deconvolution.

As a result, the following sequence is proposed by Chi, et al. in processing stacked data:

1. estimate the wavelet as well as the statistical parameters at some representative location on the line; and

2. keeping the statistical parameters constant, estimate the wavelet and reflectivity for each trace, using the wavelet estimate obtained for trace j as the initial wavelet for trace j + 1.

Initial values for a_0, b_0 and q_0 are very important for the speedy convergence of the Figure 4-20 Block Component Method. If we can determine a_0 and b_0 that provide a wavelet that is close to the true wavelet, and a q_0 that is close to the true event sequence, then this Block Component Method will converge very rapidly. We therefore replace the block labelled "Initialize Parameters," in the Figure 4-20 Block Component Method, by a second block component method, one that is expected to generate good initial values for a_0, b_0 and q_0. This new procedure is called a *two-phase block component method*; it is depicted in Figures 6-1, 6-2 and 6-3. The phase-1 block generates the initial conditions for the phase-2 block. Of course, the computational efforts for the phase-1 block must be much less than those needed for the phase-2 block; otherwise, we do not need the phase-1 block.

In the phase-1 block we use a threshold detector because it has the ability to change spikes at many locations. Because this detector is not recursive, its repeated use does not guarantee an increase in the loglikelihood function. If an increase does not occur, we abort the phase-1 block and enter the phase-2 block. We do not update the statistical parameters because Chi (1983) has shown that, if updated statistical parameters are used by the threshold detector, it will converge to very few large spikes or to no spikes at all. The phase-1 block acts as a *coarse tuning procedure*.

The phase-2 block is almost identical with the Figure 4-20 Block Component Method. It receives initial parameter values from the phase-1 block. The phase-2 block acts as a *fine tuning procedure*. Note that sub-iterations may occur within each of the major (iterative) blocks in both the phase-1 and phase-2 blocks.

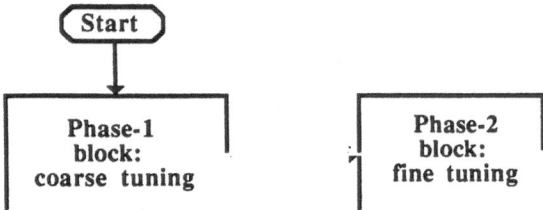

Figure 6-1. Two-Phase Block Component Method.

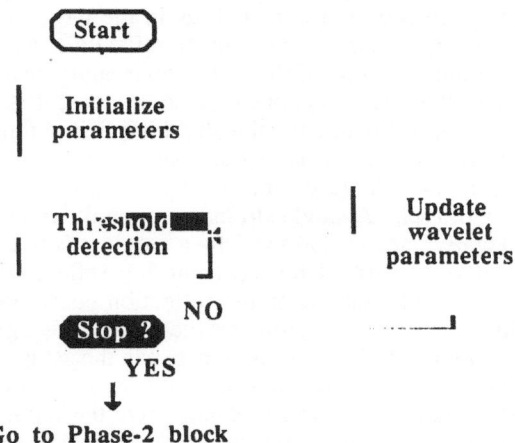

Figure 6-2. Phase-1 Block of Two-Phase Block Component Method.

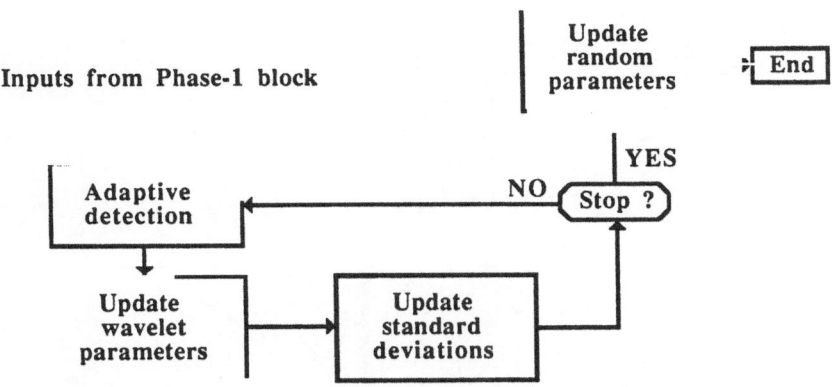

Figure 6-3. Phase-2 Block of Two-Phase Block Component Method.

Example 6-1. Our first result is depicted in Figure 6-4. The segment of stacked data which was analyzed is shown at the top of the figure, followed by the estimated reflectivity and wavelets. The reflectivity calculated from a sonic log which intersects the seismic line at approximately the first trace is also shown. The wavelet as well as statistical parameters were estimated for a single trace using the two-phase BCM initialized by a minimum-phase starting wavelet. Each successive trace was then processed through three iterations of the phase-2 block alone to yield the wavelets and reflectivity.

The estimated wavelets are very similar in character, although differences do occur in the secondary positive lobes as well as in the large negative trough, especially in the last few wavelets. The estimated reflectivity shows reasonably good continuity, especially in view of the rather poor signal level at depth. In certain locations where the reflectors appear discontinuous, it is likely that the reflection levels have dropped below the threshold implicit within the detection algorithm and a different choice of statistical parameters might improve the performance at this level. A very interesting result is the estimated reflectivity just below 0.6 sec. A single strong event on the input data has been resolved into two negative spikes followed by a single positive spike and the validity of this result is confirmed by the sonic-log reflectivity. A second interesting result is the continuous negative reflection coefficient detected at about 0.55 sec. This event is not obvious on the input data, especially on the first few traces. Unfortunately, the correlation between the estimated reflectivity and the sonic-log reflectivity is less obvious at the deeper reflections.

The estimated wavelets may be used to deconvolve the input data, and this result is shown in Figure 6-5. Minimum-variance deconvolution was used to obtain these results. One of the extracted wavelets is shown at the bottom of the figure, along with the wavelet (i.e., the MVD resolution function) after applying MVD. We see that MVD has shaped the input wavelet to an approximately zero-phase result with increased bandwidth.

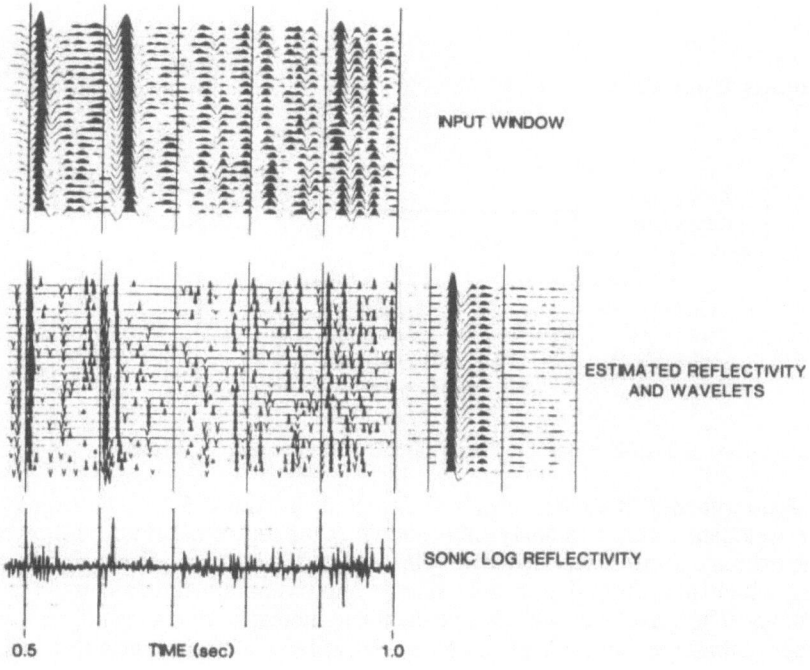

Figure 6-4.   Wavelet and reflectivity estimation. (Chi et al., 1984.   © 1984 Geophysics.)

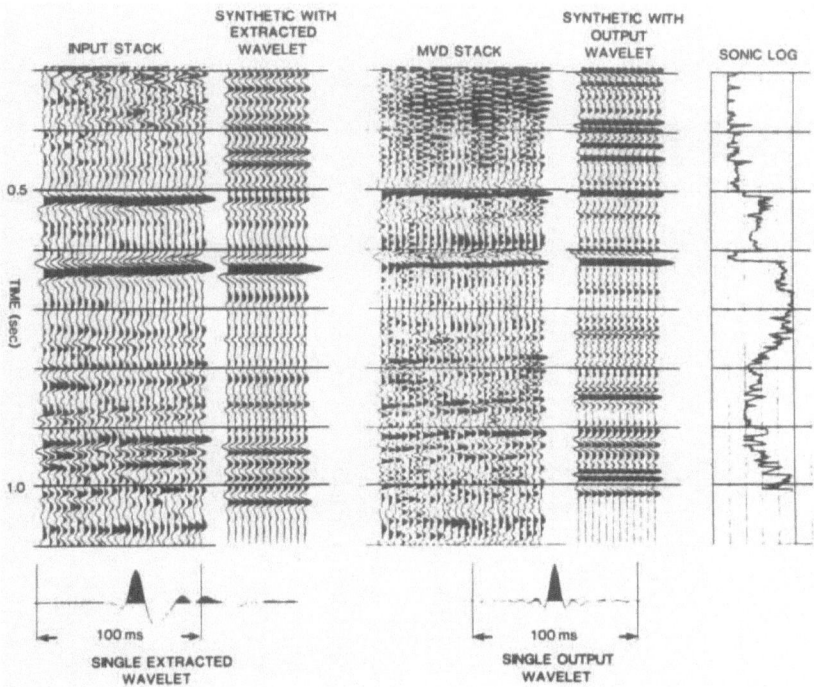

Figure 6-5. Minimum-variance deconvolution using extracted wavelets. (Chi, et al., 1984. © 1984 Geophysics.)

Synthetics were formed using the reflectivity from the sonic log and the wavelets shown. Comparing the input and output stacks, we can see a significant improvement in resolution along with a good tie to the zero-phase synthetic. The event at about 0.62 sec on the input has been resolved into two events which is confirmed by both the synthetic and the original sonic log. In general, the improved resolution at all levels is significant.

Example 6-2. Our second real data result is depicted in Figure 6-6. The exploration objective in this example is the shallow gas-sand indicated by the bright spot at 0.75 sec. Figure 6-6 shows the input window followed by the estimated reflectivity and wavelets, and finally the result of convolving the estimated reflectivity with the estimated wavelets. This last display may be compared to the input data to confirm that the estimated reflectivity and wavelets do indeed model the input data accurately. The wavelets from this example are very consistent with minor differences in the secondary lobes. At this point it is not clear whether the wavelet consistency is real or is an artifact attributable to the estimation procedure. The estimated reflectivity in Figure 6-6 shows continuous events at all the major reflectors. The gas-sand reflector at 0.75 sec on the input window now appears as a negative reflection coefficient at about 0.72 sec.

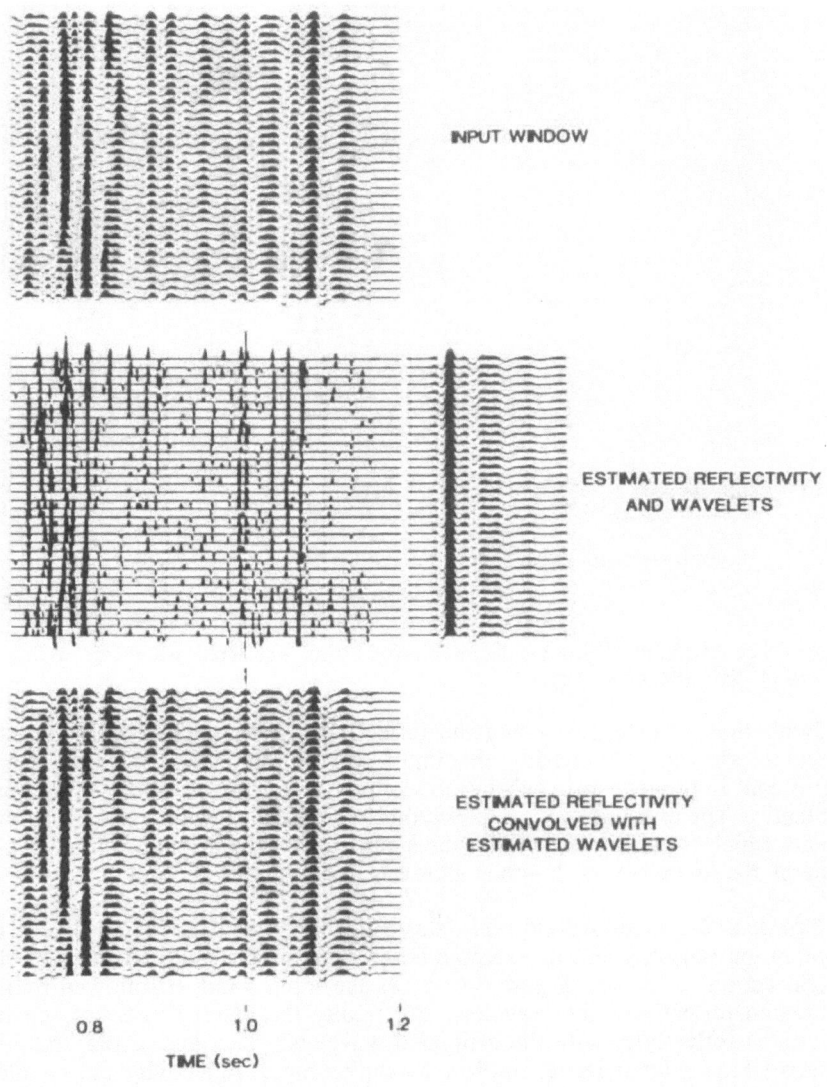

Figure 6-6.   Wavelet and reflectivity estimation. (Chi, et al., 1984. © 1984 Geophysics.)

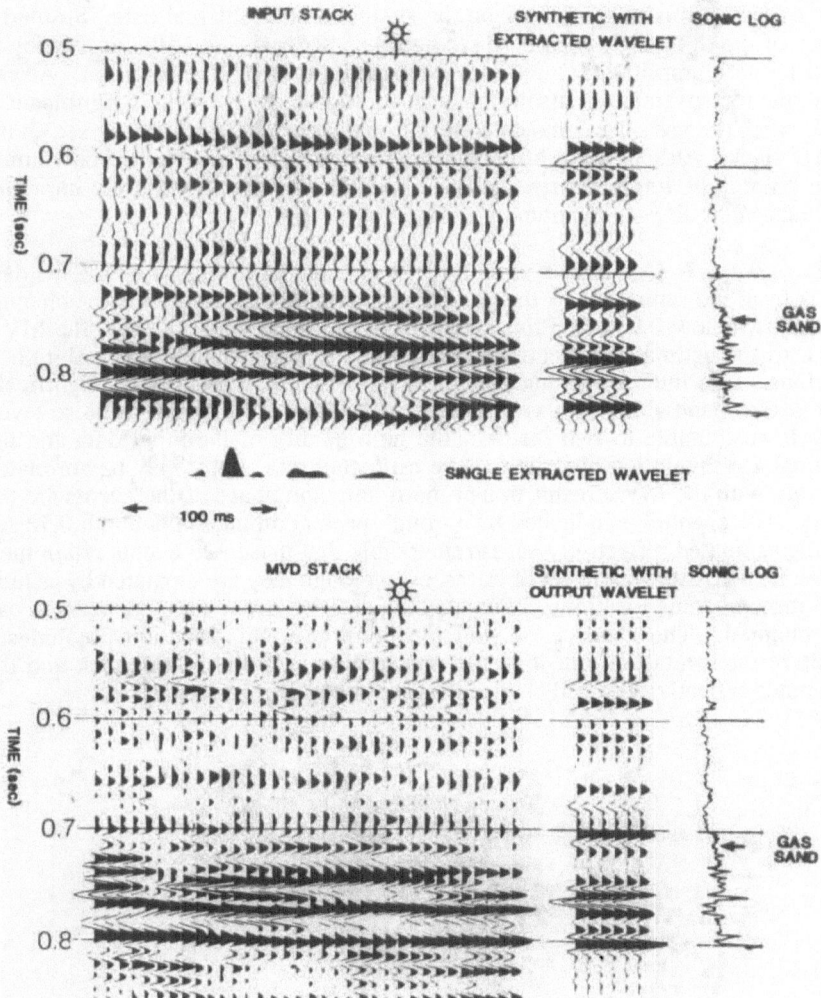

INPUT STACK · SYNTHETIC WITH EXTRACTED WAVELET · SONIC LOG

TIME (sec)

0.5
0.6
0.7
0.8

GAS SAND

SINGLE EXTRACTED WAVELET

100 ms

MVD STACK · SYNTHETIC WITH OUTPUT WAVELET · SONIC LOG

TIME (sec)

0.5
0.6
0.7
0.8

GAS SAND

SINGLE OUTPUT WAVELET

100 ms

synthetics are fairly good, the amplitude correlation in some locations is less than optimum. This would probably indicate a phase error in the extracted wavelets. The event at about 0.67 sec on the input stack, for example, shows a negative trough followed by a positive peak on both the synthetic and real data. The trough is apparantly deeper on the synthetic than the real data. Similarly, many of the events between 0.70 sec and 0.85 sec show some inconsistency in their relative amplitude, which may be indicative of a phase rotation. At any rate , the improvement in resolution as a result of deconvolution is significant at all levels. The gas sand is indicated by a trough-peak character at 0.72 sec on the MVD stack. Although the bright spot is no more obvious after deconvolution than before, the improved resolution on neighboring events certainly improves the interpretability of the anomaly.

Example 6-3. Our third real data result is depicted in Figure 6-8. This data set is from the same area as the preceding example, but the play is the channel scour just below 0.7 sec. Progressing from the input stack through the MVD stack to the estimated reflectivity, a continuous improvement in resolution is obvious. It is interesting that the MVD algorithm has produced a significant increase in bandwidth with very little degradation of the signal-to-noise level. This is attributable to two factors: the high quality of the input data and the optimal deconvolution algorithm. The estimated reflectivity may be correlated directly with the MVD result in that the timing and phase of the events are the same. The channel is indicated by a trough-peak combination at about 0.72 sec on the estimated reflectivity. It is remarkable that these two events retain their character across the entire set of traces, even though they are separated by as little as 5 msec at some locations. Note also the positive event which appears to cap the channel. The absence of well information at this location precludes a positive interpretation, but it would appear that both the MVD stack and the estimated reflectivity would be useful tools for exploration in this area.

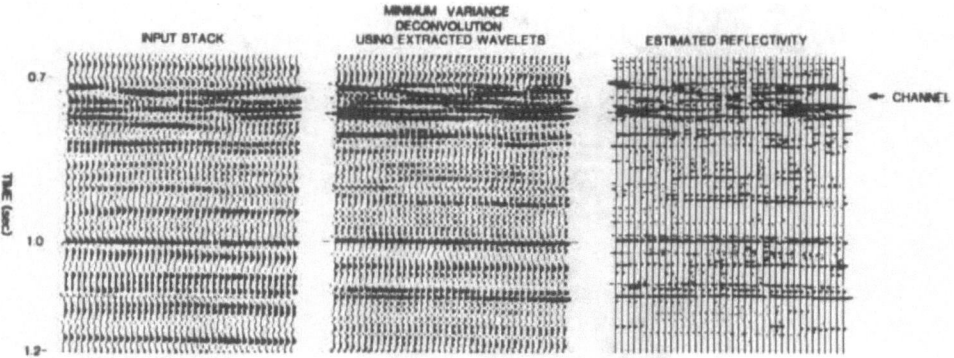

Figure 6-8. Delineation of channel scour. (Chi et al., 1984. © 1984 Geophysics.)

6.3 Minimum-Variance Deconvolution

We have repeatedly mentioned that signal-to-noise ratio and bandwidth greatly affect minimum-variance deconvolution. In this section we present two examples, each of which demonstrates these effects.

Example 6-4. To begin we describe how a synthetic seismic trace was generated. We used the reflectivity sequence depicted in Figure 6-9. For this sequence, $v_r^{1/2} = 0.15$ and $\lambda = 0.05$; hence, $v_u = v_r\lambda = 0.1125 \times 10^{-2}$. By choosing $v_r^{1/2} = 0.15$ we keep the amplitude of u(k) below 0.30, which is a geological constraint.

A fourth-order broad-band source wavelet $w_1(k)$ was used. Its transfer function is

$$W_1(z) = [- 0.76286 + 1.5884z^{-1} - 0.82356z^{-2} + 0.000222419z^{-3}]/$$
$$[1 - 2.2633z^{-1} + 1.77734z^{-2} - 0.49803z^{-3} + 0.045546z^{-4}] \quad (6\text{-}1)$$

Plots of $w_1(k)$ and its squared amplitude spectrum are depicted in Figures 6-10 and 6-11, respectively.

The broad-band wavelet was convolved with the reflectivity sequence of Figure 6-9 to obtain a noise-free seismic trace. This trace was then corrupted by additive Gaussian noise n(k), whose variance, v_n, was chosen so that signal-to-noise ratio is fixed at prespecified levels of either 10 or 5. A trace of seismic signal z(k), for signal-to-noise ratio of 10 is depicted in Figure 6-12. There is no backscatter in this data.

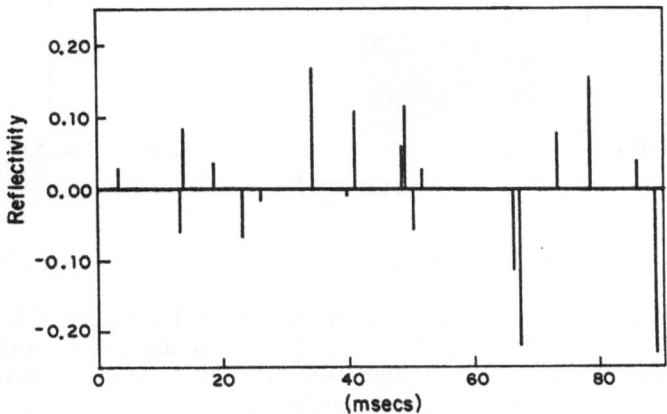

Figure 6-9. A reflectivity sequence. (Mendel, 1977. © 1977 IEEE.)

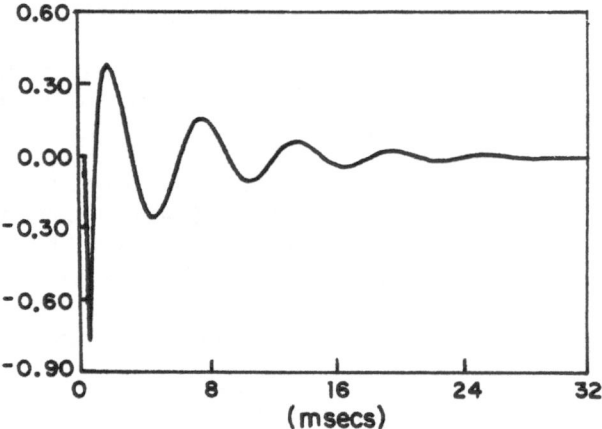

Figure 6-10.   Fourth-order broad-band wavelet.   (Kormylo and Mendel, 1980.
© 1980 IEEE.)

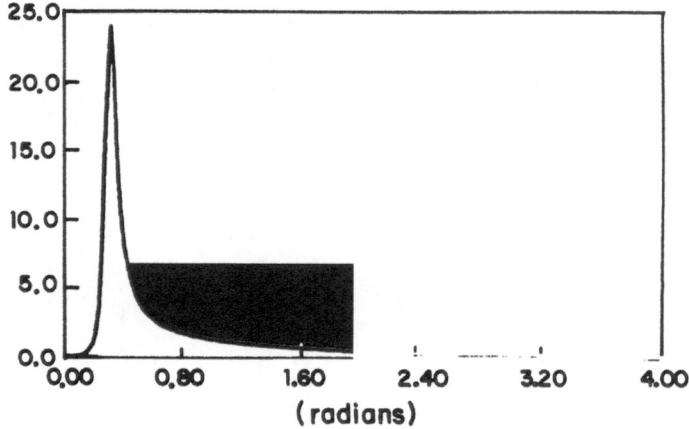

Figure 6-11. Squared amplitude spectrum of the fourth-order broad-band wavelet.

Minimum-variance deconvolution results are depicted in Figures 6-13 and 6-14. Because MVD is linear signal processing, observe that there is a value for $u^{MV}(k \mid N)$ at every sample time point. On these plots the circles denote the amplitudes of the true events in the reflectivity sequence u(k). One would hope that the MVD results produces spikes that get as close as possible to the true events. As expected, better results are obtained for the higher signal-to-noise ratio case.

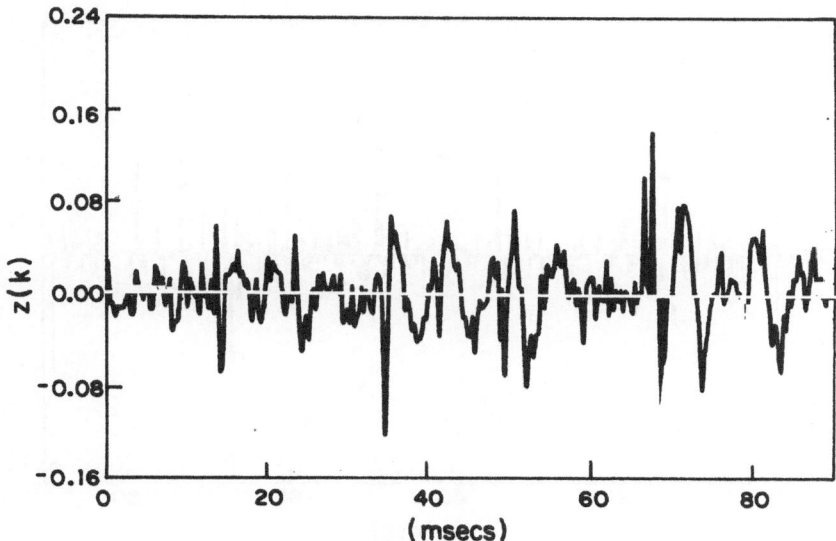

Figure 6-12. Seismic trace for signal-to-noise ratio equal to 10. (Mendel, 1977. © 1977 IEEE.)

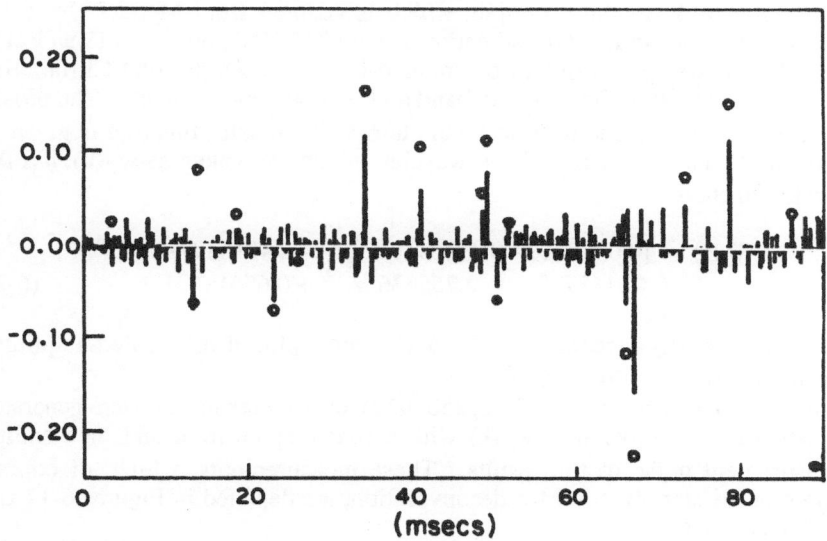

Figure 6-13. Output of MVD filter (solid lines) compared with u(k) (circles): signal-to-noise ratio equal to 10. (Mendel, 1977. © 1977 IEEE.)

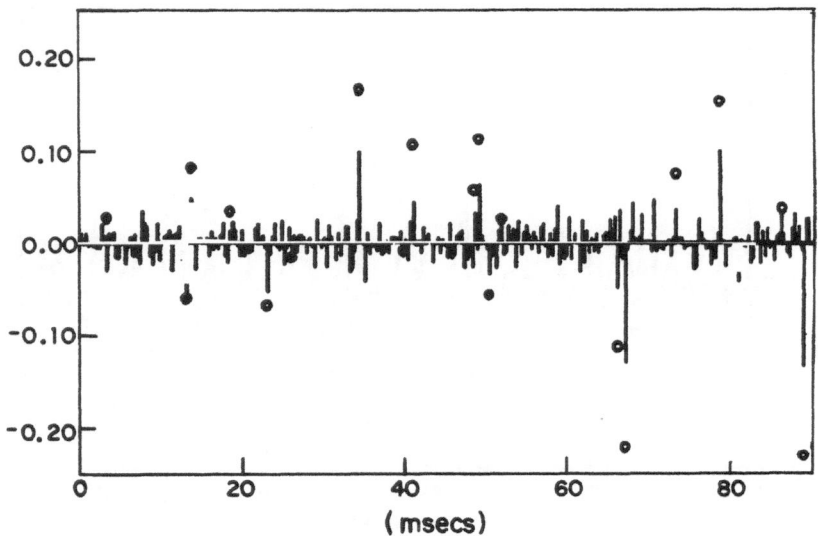

Figure 6-14.   Output of MVD filter (solid lines) compared with u(k) (circles): signal-to-noise ratio equal to 5. (Mendel, 1977. © 1977 IEEE.)

Example 6-5. In this example, which is adapted from Mendel (Jerry M. Mendel, "Lessons in Digital Estimation Theory," © 1987, pp. 207-211 and 213-214. Reprinted by permission of Prentice-Hall, Inc., Englewood Cliffs, NJ.), we compare $u^{MV}(k \mid N)$ for broad-band and narrow-band wavelets. The broad-band wavelet is the one used in Example 6-4; its transfer function is given in Equation (6-1). The narrow-band wavelet, which we denote as $w_2(k)$, has the transfer function

$$W_2(z) = [0.0378417 - 0.0306517z^{-2}]/[1 - 3.4016497z^{-1} + 4.5113732z^{-2} - 2.7553363z^{-3} + 0.6561z^{-4}]. \tag{6-2}$$

A plot of $w_2(k)$ appears in Figure 6-15, and, a plot of its squared amplitude spectrum appears in Figure 6-16.

Measurements, $z(k)$ ($k = 1, 2, ..., 250$, taken every three msec), were generated by convolving $w_1(k)$ and $w_2(k)$ with a sparse spike train and then adding measurement noise to the results. These measurements, which, of course, represent the starting point for deconvolution, are depicted in Figures 6-17 and 6-18, respectively.

The MVD results are depicted in Figures 6-19 and 6-20. Observe that much better results are obtained for the broad-band wavelet than for the narrow-band wavelet, even though data quality, as measured by signal-to-noise ratio is much lower in the former case. The MVD results for the narrow-band wavelet appear smeared out, whereas the MVD results for the broad-band wavelet are quite sharp. This smearing effect can be explained by examining the *resolution functions* for these two cases.

Resolution functions $\rho_1(k)$ and $\rho_2(k)$, associated with wavelets $w_1(k)$ and $w_2(k)$, respectively, are depicted in Figures 6-21 and 6-22. As predicted by Equation (5-7), $\rho_1(k)$ is much spikier than $\rho_2(k)$, which explains why the MVD results for the broad-band wavelet are quite sharp, whereas the MVD results for the narrow-band wavelet are smeared out.

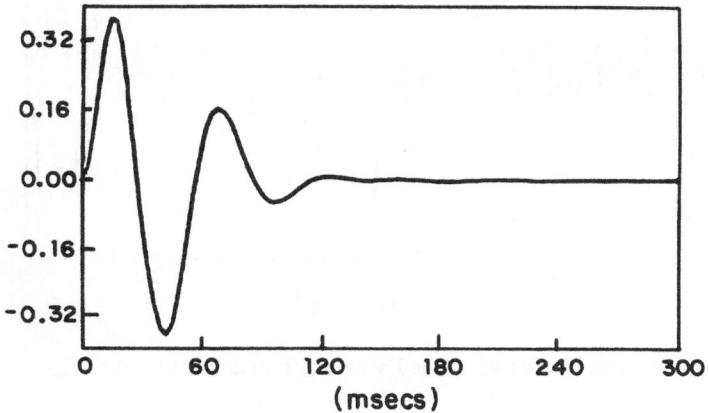

Figure 6-15. Fourth-order narrow-band wavelet. (Chi and Mendel, 1984. © 1984, IEEE.)

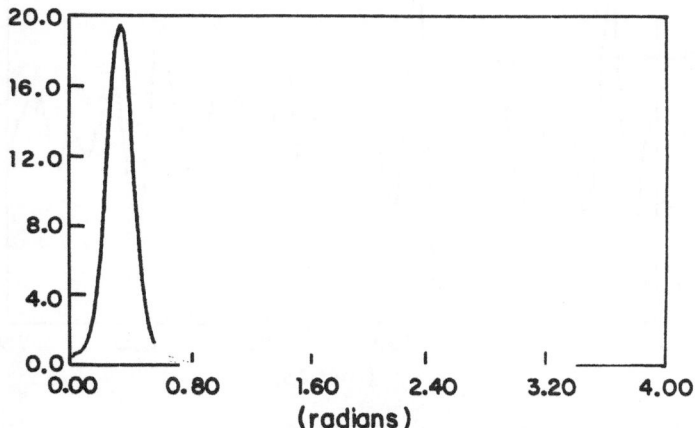

Figure 6-16. Squared amplitude spectrum of the fourth-order narrow-band wavelet. (Chi and Mendel, 1984. © 1984, IEEE.)

Observe, also, that $u^{MV}(k \mid N)$ tends to undershoot $u(k)$. This can be explained by comparing the peak amplitudes of $\rho_1(k)$ and $\rho_2(k)$. Because the peak amplitude of $\rho_2(k)$ is only about 0.32, whereas the peak amplitude of $\rho_1(k)$ is

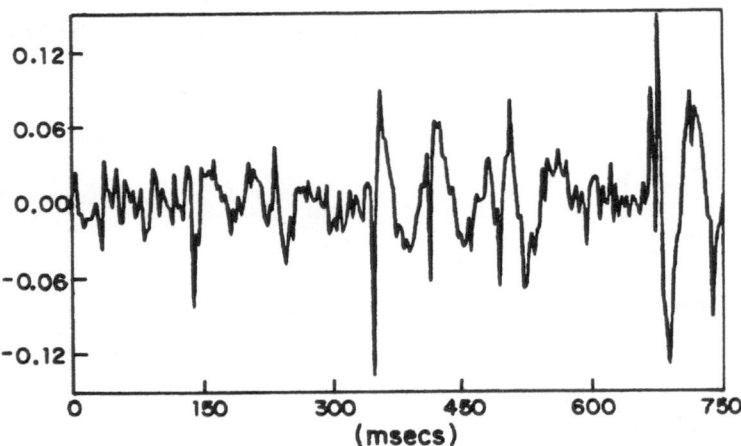

Figure 6-17. Measurements associated with the broadband wavelet; signal-to-noise ratio equal to 10. (Chi, 1983).

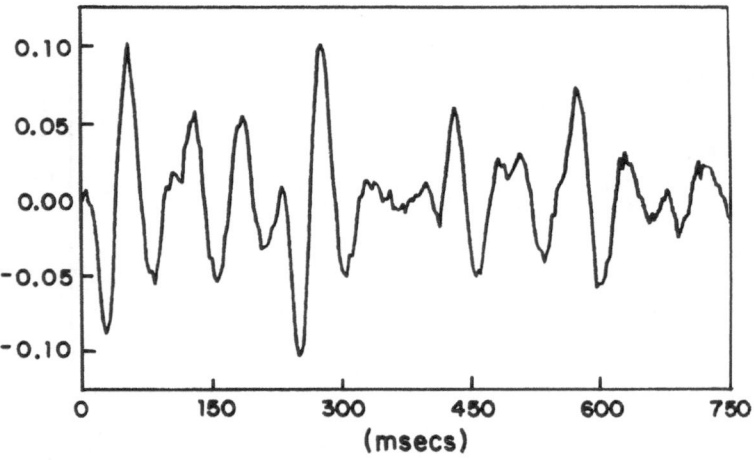

Figure 6-18. Measurements associated with the narrow-band wavelet; signal-to-noise ratio equal to 100. (Chi and Mendel, 1984b, © 1984, IEEE).

about 0.80, we see now why the clumps of spikes in Figure 6-20 are so much lower than the spikes in Figure 6-19.

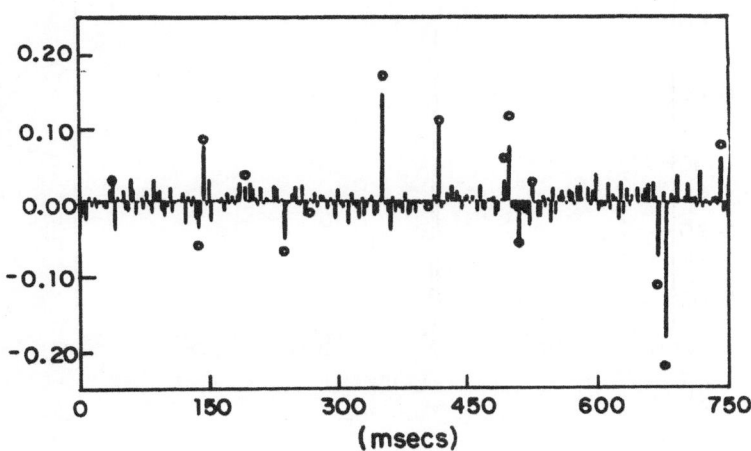

Figure 6-19. $u^{MV}(k \mid N)$ for the broad-band wavelet and the data depicted in Figure 6-17. (Chi, 1983).

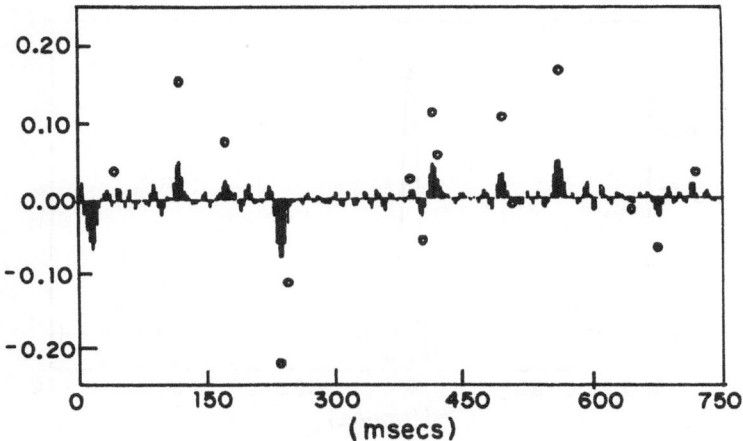

Figure 6-20. $u^{MV}(k \mid N)$ for the narrow-band wavelet and the data depicted in Figure 6-18. (Chi and Mendel, 1984b, © 1984 IEEE).

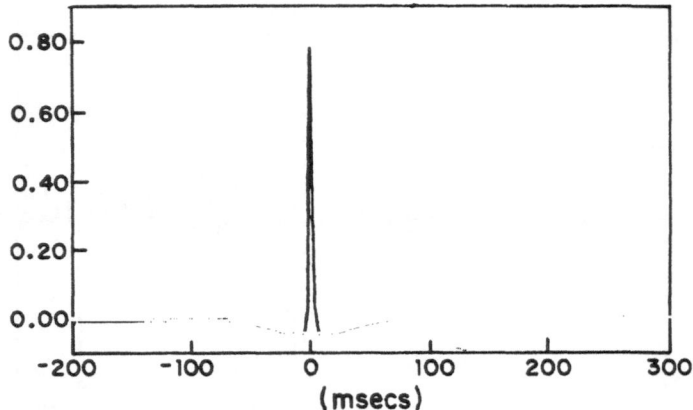

Figure 6-21. Resolution function for the broad-band wavelet and signal-to-noise ratio equal to 10. (Chi and Mendel, 1984b, © 1984, IEEE).

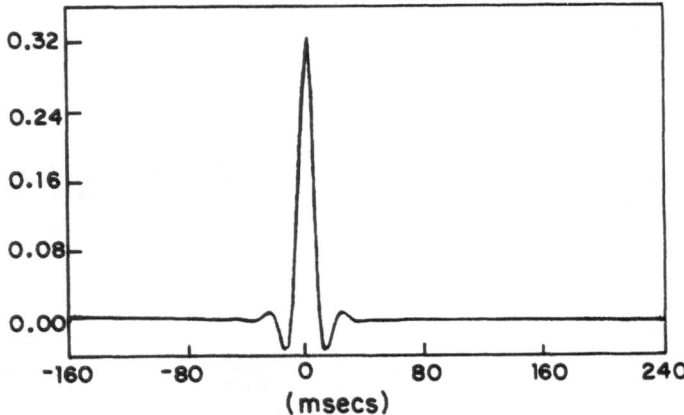

Figure 6-22. Resolution function for the narrow-band wavelet and signal-to-noise ratio equal to 100. (Chi and Mendel, 1984b, © 1984, IEEE).

6.4 Detection

In Chapter 4 we described a variety of detectors, ranging from the one-shot threshold detector to the recursive SMLR, SSS, SMLR-SSS, and MMLR detectors. In this section we present four examples that illustrate many of the points that were made about these detectors either in Chapter 4 or in Chapter 5.

Example 6-6. In Chapter 5, in the section entitled "Detectors," we stated that the probability of detection, $p_d(k)$, is much larger than the probability of false alarm, $p_f(k)$, for a threshold detector. In this example, which is taken from Shiva (1982), we compare these two probabilities for the broad-band wavelet $w_1(k)$, which was introduced in Example 6-4, and is depicted in Figure 6-10. As in Example 6-4, we assume that $v_r^{1/2} = 0.15$ and $\lambda = 0.05$.

Figure 6-23 depicts the ratio of probability of detection to probability of false alarm for signal-to-noise ratios of 20, 10 and 5. The plots were obtained using the formulas for these probabilities that are given in Equations (8-22) and (8-21). Observe that $p_d(k)/p_f(k)$ is very large and decreases as signal-to-noise ratio decreases.

This example demonstrates that using the threshold detector we can have a very high probability of detecting the right spikes. We expect poorer results for narrower-band wavelets, such as the one depicted in Figure 6-15, and, performance deteriorates as signal-to-noise ratio decreases.

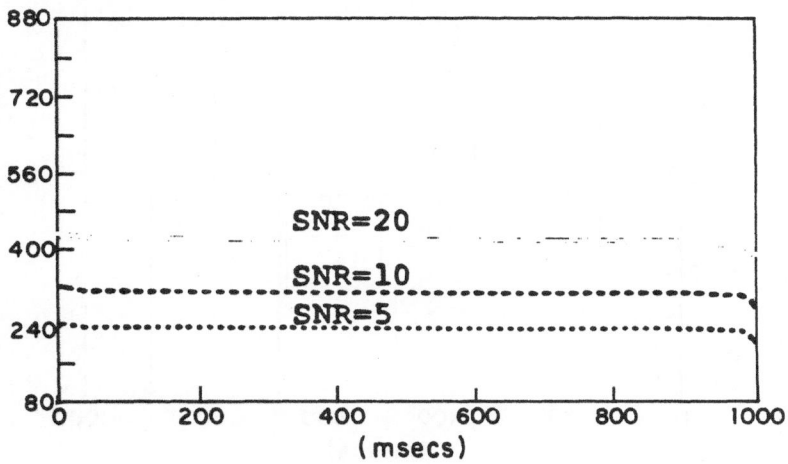

Figure 6-23. $p_d(k)/p_f(k)$ for signal-to-noise ratios (SNR) of 20, 10 and 5. (Shiva, 1982.)

Example 6-7.   This example compares the results from two SMLR detectors One is based on the objective function $\mathcal{M}$ and the other is based on the loglikelihood function $\mathcal{N}$. Recall, from Figure 5-3, and its accompanying discussions, that the former SMLR detector is supposed to produce too many false alarms, whereas the latter SMLR detector is supposed to be well behaved. Once again, our results are for the Figure 6-10 broad-band wavelet. Results from the two detectors are depicted in Figures 6-24 and 6-25, respectively. Observe that the results from the detector that is based on function $\mathcal{M}$ does indeed produce alot of false alarms, whereas the detector that is based on the loglikelihood function $\mathcal{N}$ does not. Hence this example supports the use of the latter detector.

Remember our conjecture, though, that, if backscatter is included in the model, then the SMLR detector that is based on $\mathcal{M}$ should work just fine. This is borne out by the results that are depicted in Figure 6-26.

Example 6-8.   We have observed that spike splitting and shifting occur when the SMLR detector is used and the wavelet is not broad-band. Spike-splitting can also occur for broad-band wavelets that have a slowly rising onset. This example, which is taken from Chi and Mendel (1984a), demonstrates that both the SSS and SSS-SMLR detectors can help to improve the detection results obtained by the SMLR detector.

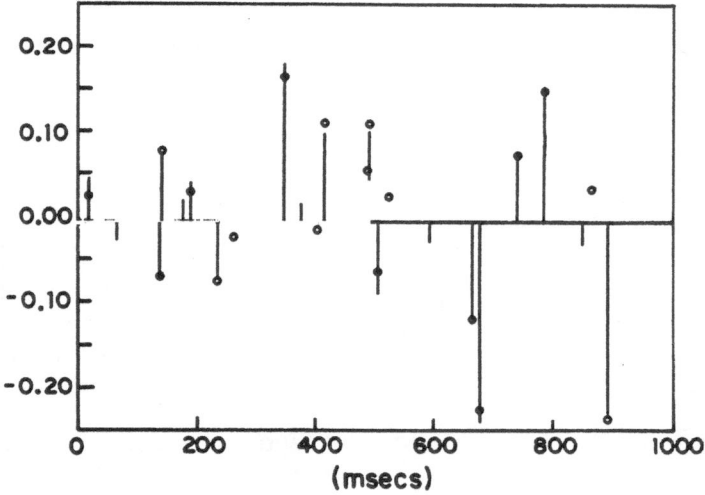

Figure 6-24. Estimated reflections from the SMLR detector that is based on the objective function $\mathcal{M}$.

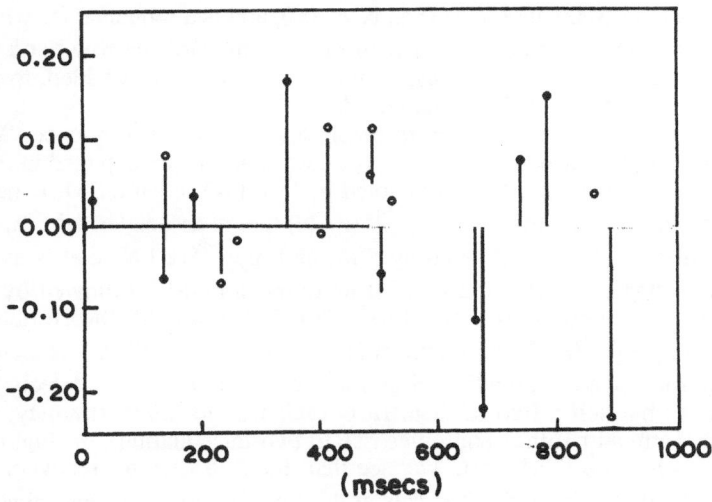

Figure 6-25. Estimated reflections from the SMLR detector that is based on the loglikelihood function $\mathcal{N}$.

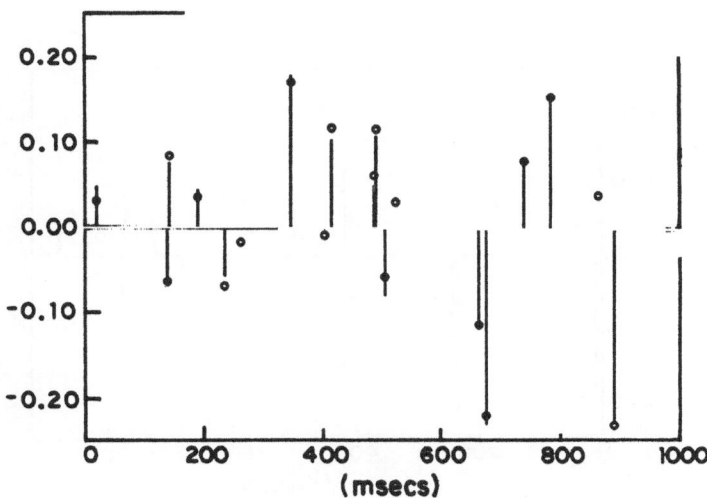

Figure 6-26. Estimated reflections from the SMLR detector that is based on the objective function $\mathcal{M}$ when backscatter is included.

The reflectivity sequence that is depicted in Figure 6-27 was used. It has 18 spikes, contains 300 data points (i.e., N = 300), and is a sequence for which λ = 0.05. This signal was convolved with the fourth-order narrow-band wavelet depicted in the insert in Fig. 6-28, to which white noise was added, to produce the synthetic data depicted in Figure 6-28.

A threshold detector was used to obtain an initial event sequence. We then examined the three schemes for recursive detection that are depicted in Figure 6-29. Scheme 1, whose output is denoted $u_1^{MV}(k \mid N)$, is pure SMLR detection. Scheme 2 has two components: $u_{21}^{MV}(k \mid N)$ that is associated with two stages of detection --- SMLR followed by SSS; and, $u_{22}^{MV}(k \mid N)$ that is associated with three stages of detection --- SMLR followed by SSS followed by SMLR. Scheme 3, whose output is denoted $u_3^{MV}(k \mid N)$, is pure SSS-SMLR detection.

Observing $u_1^{MV}(k \mid N)$ in Figure 6-30, we see that the SMLR detector splits the first true spike (i.e., the first circle) into two smaller spikes, detects 11 other true spikes but shifts five of them from their true locations (namely, the 6th, 8th, 9th, 14th, and 16th) , and gives rise to two false alarms. In Figure 6-31, which depicts $u_{21}^{MV}(k \mid N)$, we see that the SSS detector recovers the first spike, detects the 8th and 9th spikes at their correct locations and gives rise to only one false alarm. Only three detected spikes in $u_{21}^{MV}(k \mid N)$ are shifted from their true locations, namely the 6th, 14th and 16th. In Figure 6-32, which depicts $u_{22}^{MV}(k \mid N)$, we see that the cascaded SSS and SMLR detectors eliminate false alarms; but the 6th, 14th and 16th detected spikes still remain shifted. In Figure 6-33, which depicts $u_3^{MV}(k \mid N)$, we see that the SSS-SMLR detector has shifted the 6th spike to its correct location, and only the 14th and 16th detected spikes remain shifted from their true locations.

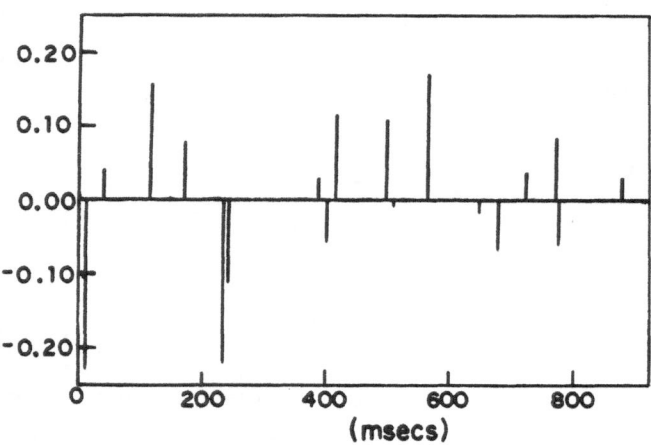

Figure 6-27. Another reflectivity sequence. (Chi and Mendel, 1984a. © 1984 IEEE.)

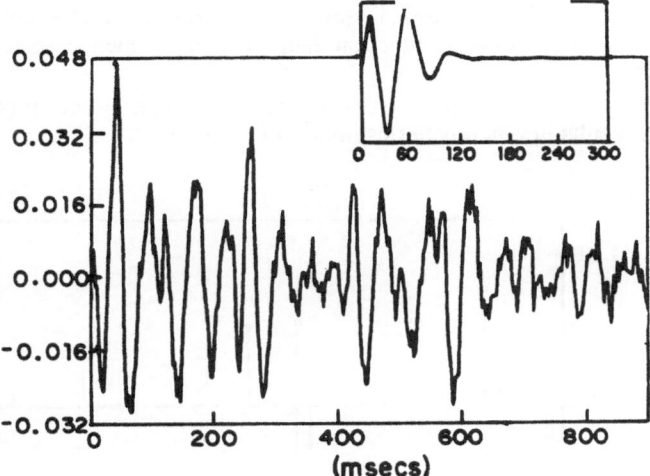

Figure 6-28. Synthetic data for which signal-to-noise ratio equals 25. (Modified from Chi and Mendel, 1984a. © 1984 IEEE.) Insert is of a fourth-order narrowband wavelet.

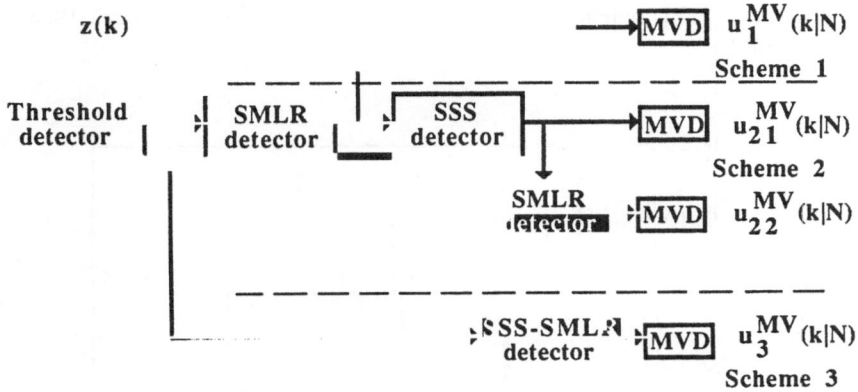

Figure 6-29. Simulation schemes. (Chi and Mendel, 1984a. © 1984 IEEE.)

From these results, we see that the SSS detector can improve SMLR detector results, and vice versa when they are cascaded together, and, the SSS-SMLR detector works even better. Another experiment was performed in which $u_{22}{}^{MV}(k \mid N)$ was used as the starting sequence for a second cascade of SSS and SMLR detectors. We were able to improve $u_{22}{}^{MV}(k \mid N)$, and, in fact we obtained the same sequence as $u_3{}^{MV}(k \mid N)$. Doing this again, we did not improve our results. Apparantly, $u_3^{MV}(k \mid N)$ is a locally optimal estimate of $u(k)$ and neither the SSS- or SMLR-detectors can find another event sequence that has a higher loglikelihood function than the event sequence associated with $u_3^{MV}(k \mid N)$.

Finally, observe from Figures 6-30 through 6-33, that better spike location information results in much better estimates of amplitudes.

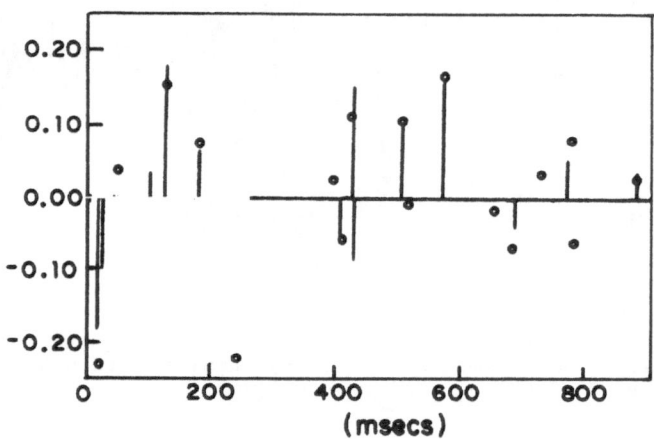

Figure 6-30. MVD output $u_1^{MV}(k \mid N)$ from scheme 1. (Chi and Mendel, 1984a. © 1984 IEEE.)

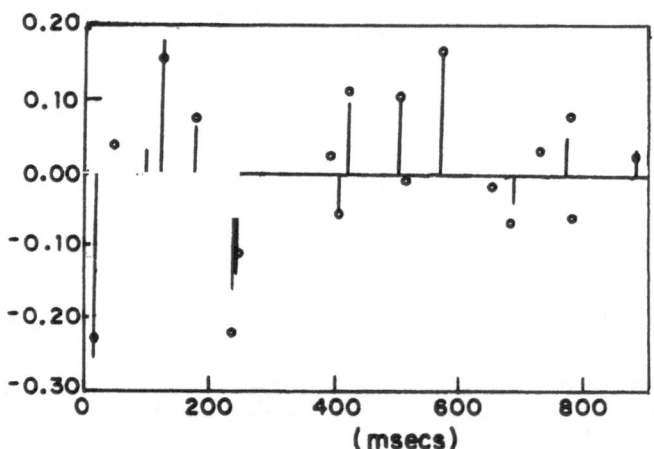

Figure 6-31. MVD output $u_{21}^{MV}(k \mid N)$ from scheme 2. (Chi and Mendel, 1984a. © 1984 IEEE.)

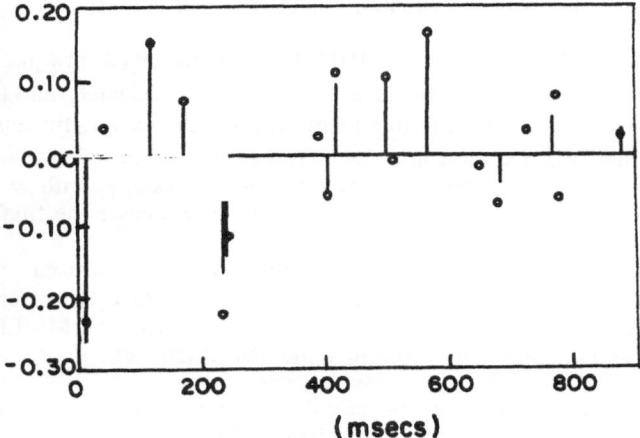

Figure 6-32. MVD output $u_{22}^{MV}(k \mid N)$ from scheme 2. (Chi and Mendel, 1984a. © 1984 IEEE.)

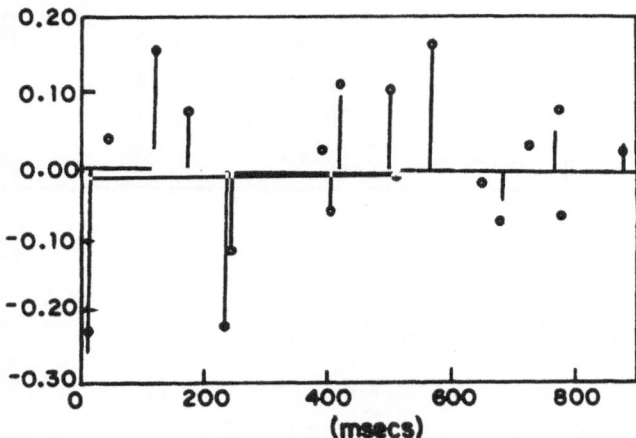

Figure 6-33. MVD output $u_3^{MV}(k \mid N)$ from scheme 3. (Chi and Mendel, 1984a. © 1984 IEEE.)

Example 6-9. This example, which is our last one on detection, illustrates the computational advantage of the MMLR detector over the SMLR detector. It is taken from Chi, et al. (1985). Using a pseudo-random number generator, they generated a 200 point Bernoulli-Gaussian sequence for u(k) with nonzero values:

$$u(2) = 0.3023760, \quad u(50) = 0.2915920, \quad u(58) = -0.1446380,$$
$$u(102) = 0.1288300, \quad u(150) = -0.1447480.$$

For this sequence $\lambda = 0.025$, $v_r = 0.04$, N = 200 and there are just 5 spikes. This spike sequence was convolved with a third-order wavelet, and then noise ($v_n = 0.15 \times 10^{-2}$) was added to this result to produce a synthetic seismogram. The SMLR and MMLR detectors were run on this data using a number of different initial event sequences. A fixed window strategy, with w = 20 (see Figure 4-13), was used for the MMLR detector. In all cases the final detected event sequence was the same.

Table 6-1 summarizes four different cases. It shows the total number of iterations required for both detectors to converge, together with associated CPU times (the simulations were performed on a VAX 11/750). The MMLR detector is in all cases computationally faster than the SMLR detector.

Table 6-1. Number of iterations for convergence and CPU times required by the SMLR and MMLR detectors. (Chi, et al. 1985. © 1985 IEEE.)

Initial Event Sequence: Number of Spikes	SMLR Detector		MMLR Detector	
	Number of Iterations for Convergence	CPU Time (sec)	Number of Iterations for Convergence	CPU Time (sec)
49	40	64.91	8	15.98
34	25	40.30	4	8.13
24	17	28.46	4	8.42
15	12	19.13	3	7.63

6.5 Block Component Method

In this section we present some examples that illustrate different aspects of a block component method.

Example 6-10. This example, which illustrates the two-phase *block component method* that was described at the beginning of this chapter, is taken from Chi, et al. (1984). The seismic wavelet is the one that is depicted in Figure 6-10; the reflectivity is depicted in Figure 6-9; and, the synthetic seismogram is depicted in Figure 6-12.

The Phase-1 block of the two-phase block component method is depicted in Figure 6-2. For the data of this example, 5 iterations were spent in the Phase-1 block. Figures 6-34 and 6-35 depict the estimated wavelet and reflections from this block. Observe that the estimated wavelet is quite close to the actual wavelet, and that the larger spikes have all been detected by the threshold detector. Very good initial conditions have been generated for the Phase-2 block.

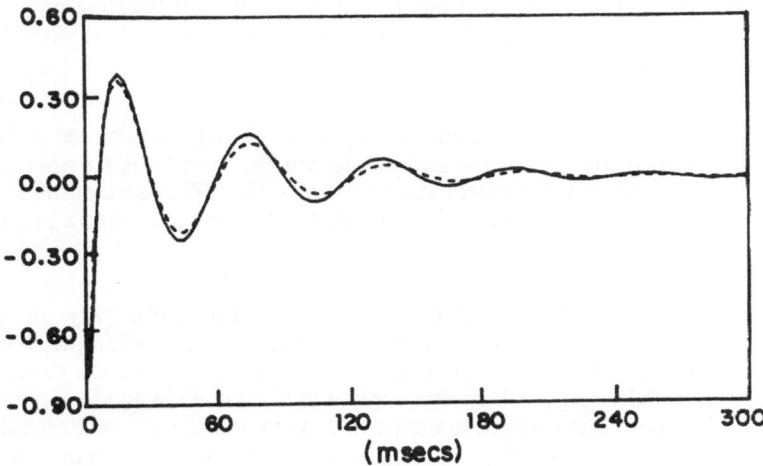

Figure 6-34.   Estimated wavelet from Phase-1 block.   Solid line depicts true wavelet and dashed line depicts the estimated wavelet. (Chi et al., 1984. © 1984 Geophysics.)

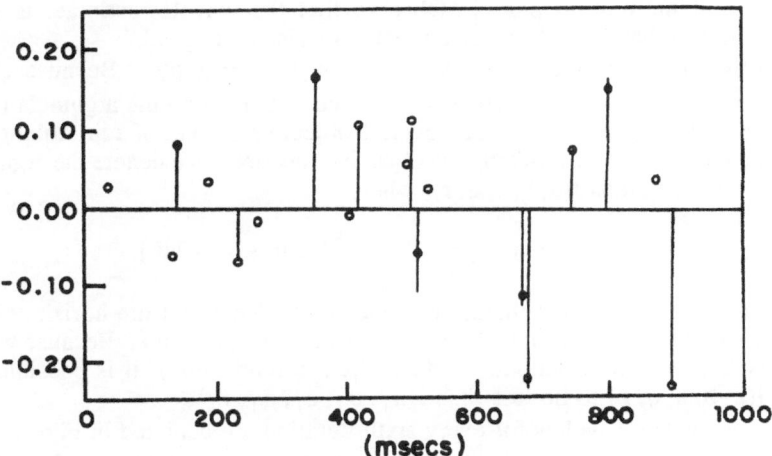

Figure 6-35. Estimated reflections from Phase-1 block. (Chi et al., 1984. ©1984 Geophysics.)

Six iterations were spent in the Phase 2 block of the two-phase block component method. The final estimated wavelet is nonminimum phase; its transfer function is (see Eq.(6-1))

$$W_1^{ML}(z) = [-1.197387 + 2.413965z^{-1} - 1.113551z^{-2} - 0.102151z^{-3}]/$$
$$[1 - 2.183021z^{-1} + 1.591139z^{-2} - 0.361159z^{-3}$$
$$+ 0.012296z^{-4}] . \qquad (6-3)$$

The scaled estimated wavelet and reflectivity sequence are depicted in Figures 6-36 and 6-37, respectively (the scale factor for the wavelet is C, whereas it is 1/C for the reflectivity sequence, and C = 0.637104). We see that the Phase-2 block has accomplished its objective, and that it does indeed act as a fine tuning procedure.

Example 6-11. There appears to be a predictable way in which convergence occurs in a Block Component method. This example, which is taken from Mendel (1983), illustrates this behavior. The Block Component Method depicted in Figure 4-20 was applied to the data depicted in Figure 6-12. After 25 iterations it converged to the wavelet and reflection estimates depicted in Figures 6-38 and 6-39, respectively. The estimated values for λ and v_n are 0.05 and 1.66×10^{-2}, respectively. The z-transform transfer function for the estimated wavelet is

$$W(z) = 0.55453 \ (z^{-1} - 0.90613)(z^{-1} - 1.0150)(z^{-1} - 24.014)/$$
$$(z^{-1} - 1.0022 \pm 0.32001j)(z^{-1} - 3.4719 \pm 1.4926j), \qquad (6-4)$$

which is nonminimum phase. Observe that the wavelet estimate is quite excellent, and that the reflectivity estimates are also quite good.

Loglikelihood function $\mathcal{N}$ was used in this example. Because $\mathcal{N}$ is exponential in nature, it is convenient to focus attention on the argument of the exponential, and to rephrase our overall objective as one of minimizing this argument instead of maximizing the objective function. We denote the argument of the exponential as $J(\mathbf{a}, \mathbf{b}, \mathbf{s}, \mathbf{q}, \mathbf{r})$, where

$$J(\mathbf{a}, \mathbf{b}, \mathbf{s}, \mathbf{q}, \mathbf{r}) = -2 \ \ln \mathcal{N} \ \{\mathbf{a}, \mathbf{b}, \mathbf{s}, \mathbf{q}, \mathbf{r} \mid \mathbf{z} \} \qquad (6-5)$$

In Figure 6-40 we depict the convergence of J. Note that the horizontal line marks the value of J given the "true" values of $\mathbf{a}$, $\mathbf{b}$, $\mathbf{s}$, $\mathbf{q}$, and $\mathbf{r}$. Because we are processing a finite amount of data (i.e., a realization) it is possible for min $J(\mathbf{a}, \mathbf{b}, \mathbf{s}, \mathbf{q}, \mathbf{r})$ to be less than $J(\mathbf{a}_T, \mathbf{b}_T, \mathbf{s}_T, \mathbf{q}_T, \mathbf{r}_T)$.

The estimated wavelets for every sixth iteration are depicted in Figure 6-41. Additionally convergence of λ^{ML} and v_n^{ML} is depicted in Figures 6-42 and 6-43, respectively. In these figures, the horizontal lines mark the "true" values of λ and v_n, respectively.

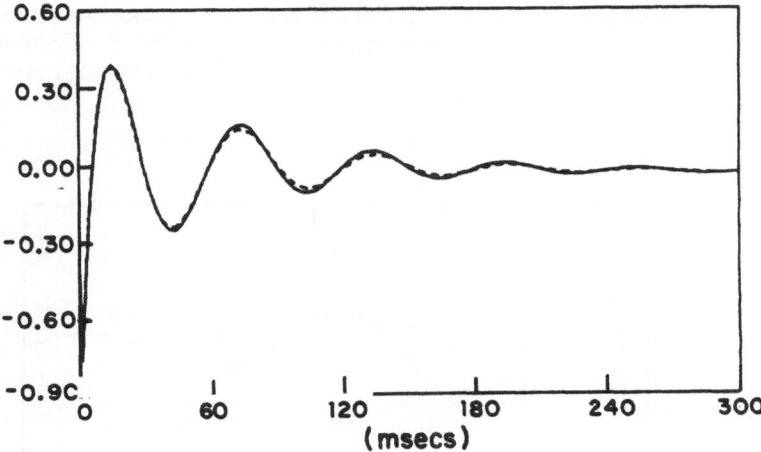

Figure 6-36. Estimated wavelet from   Phase-2 block. Solid line depicts true wavelet and dashed line depicts the estimated wavelet. (Chi et al., 1984. © 1984 Geophysics.)

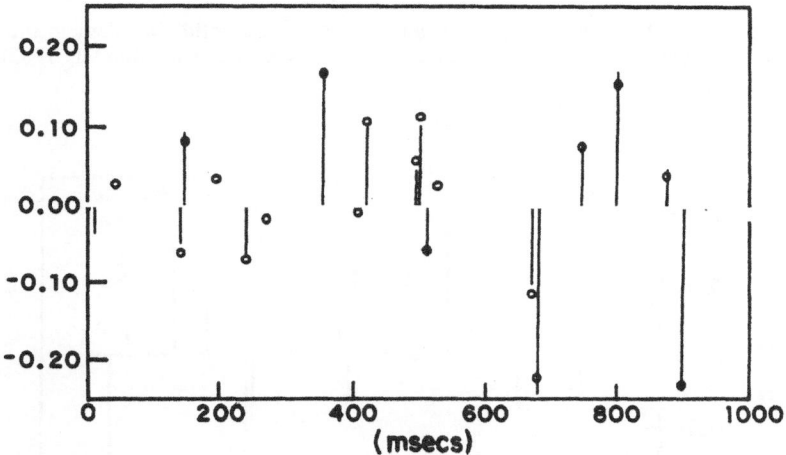

Figure 6-37. Estimated reflections from Phase-2 block. (Chi et al., 1984. © 1984 Geophysics.)

Kormylo (1979) noted, that convergence of our block component method seems to occur in three stages. The first stage (iterations 1-4) consists of convergence to an incorrect phase realization of the wavelet and is characterized by a rapid increase in $\mathscr{N}$ (or, decrease in J) (Figure 6-40). The modeling errors for these early wavelets produce large values for v_n^{ML} (Figure 6-43). Because using an incorrect phase realization for the wavelet corresponds to convolving the reflection signal with an all-pass filter, we detect 'bunches' of reflections for each significant true reflection, as shown in Figure 6-44. This also produces large values for λ^{ML} (Figure 6-42).

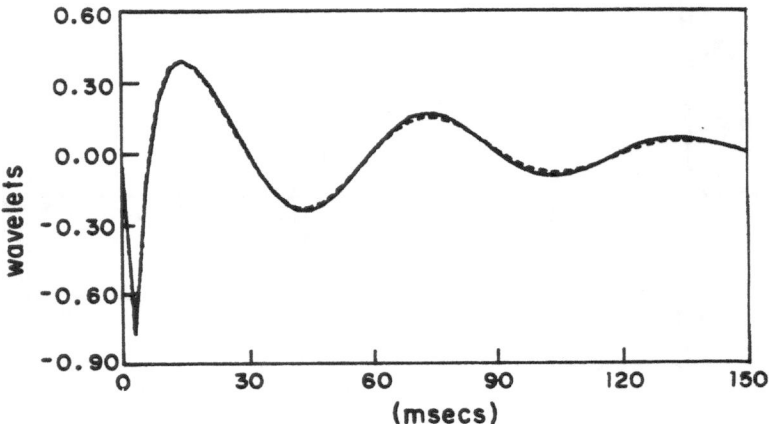

Figure 6-38.   Estimated ARMA wavelet (n = 4).   The solid line depicts the true wavelet, the dashed line the estimated wavelet. (Kormylo and Mendel, 1983b. © 1983 IEEE.)

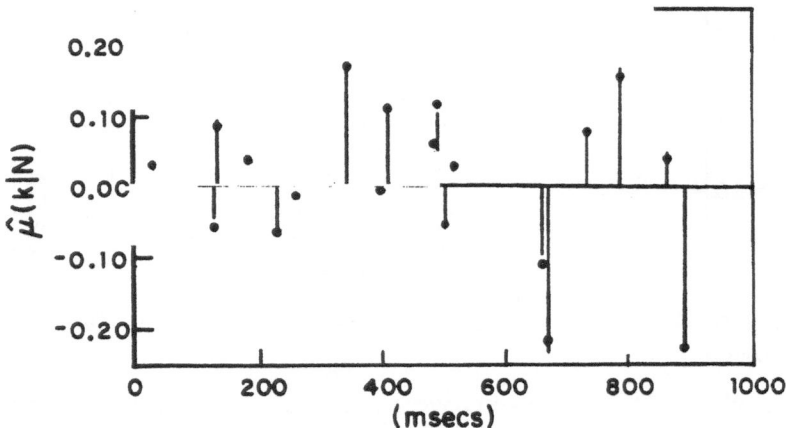

Figure 6-39.   Estimated reflections.   Circles depict true reflections, bars estimated reflections.   (Kormylo and Mendel, 1983b. © 1983 IEEE.)

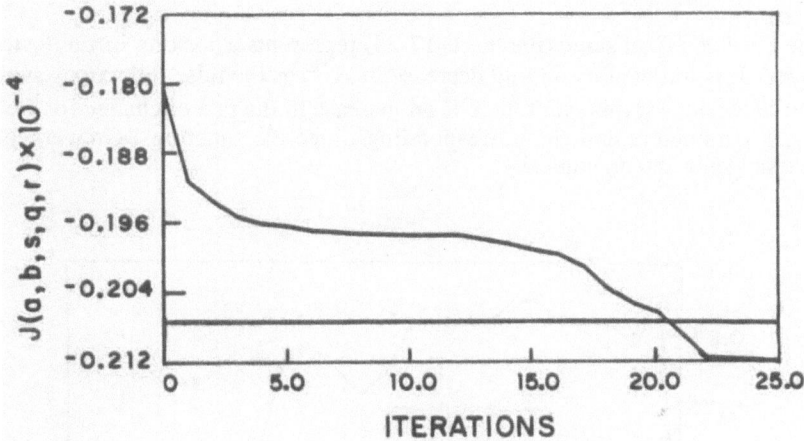

Figure 6-40. Convergence of $J(a, b, s, q, r)$. The horizontal line marks $J(a_T, b_T, s_T, q_T, r_T)$. (Kormylo and Mendel, 1983b. © 1983 IEEE.)

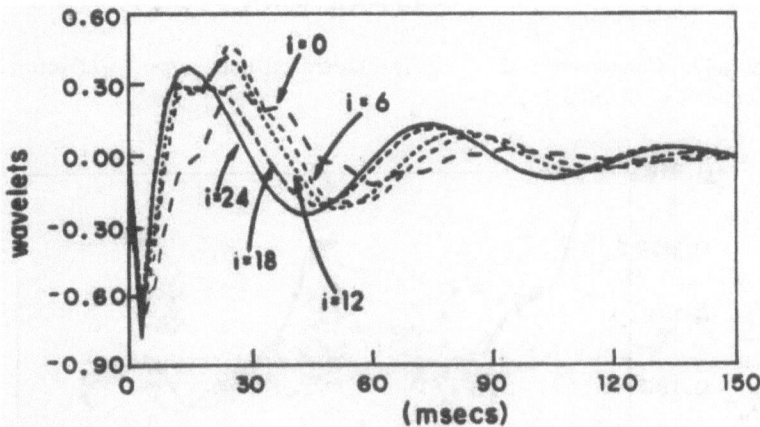

Figure 6-41. Convergence of wavelet estimates. (Kormylo and Mendel, 1983b. © 1983 IEEE.)

The second stage (iterations 5-16) consists of converging to the correct phase realization for the wavelet, and is characterized by a much slower increase in $\mathcal{N}$. During this stage the detected bunches of reflections become more like the single reflectors, with the relative amplitudes of false reflections to true reflections gradually decreasing until finally the false reflections are no longer detected.

The third and final stage (iterations 17-25) represents a 'locking in' on the true wavelet. It is initiated by a rapid decrease in λ^{ML} as the false reflections are no longer detected. At this point there is an increase in the rate of change for the **a**, **b**, and **s** parameters and the corresponding objective function. Convergence occurs suddenly and dramatically.

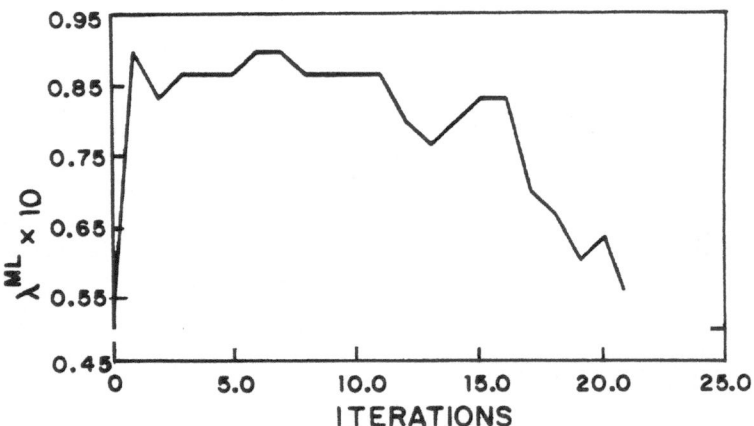

Figure 6-42. Convergence of λ^{ML}. The horizontal line marks λ_T. (Kormylo and Mendel, 1983b. © 1983 IEEE.)

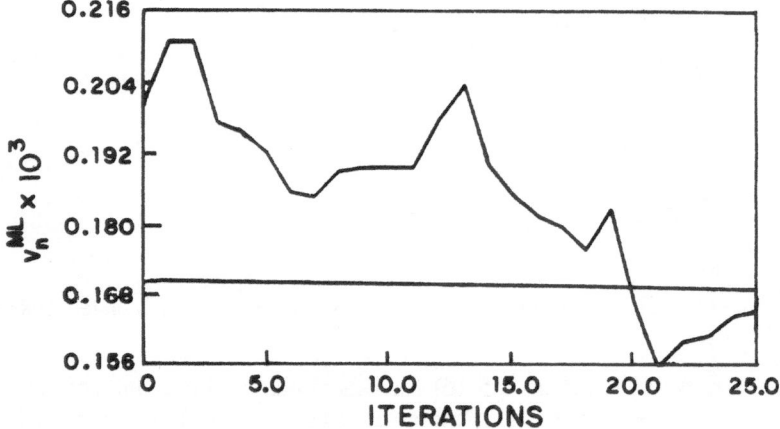

Figure 6-43. Convergence of $v_n{}^{ML}$. The horizontal line marks the true value of v_n. (Kormylo and Mendel, 1983b. © 1983 IEEE.)

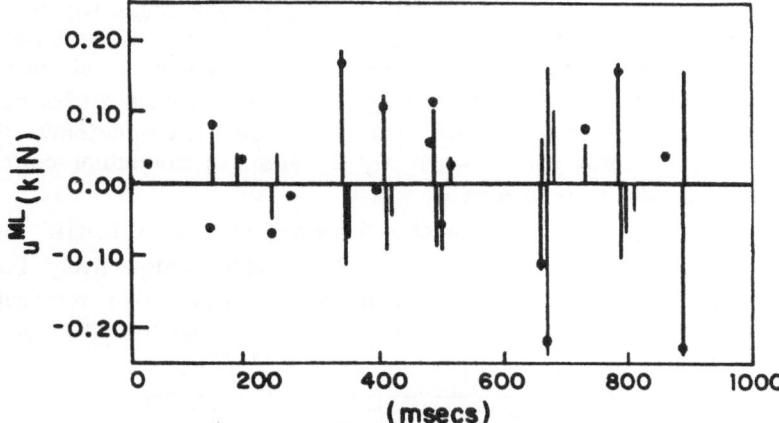

Figure 6-44.   Estimated reflections after four iterations.   Circles depict true reflections, bars estimated reflections. (Kormylo and Mendel, 1983b. © 1983 IEEE.)

A word of caution is in order. When narrower bandwidth wavelets are used, then, in the second stage of convergence, false reflections are *not* entirely eliminated. They cannot be because of bandwidth considerations that we have discussed in Chapter 5. Convergence to the correct phase realization   for   the wavelet   still   occurs; but, reflectivity estimates converge to sequences that do have false alarms.

6.6 Backscatter

The importance of backscatter has been discussed in previous chapters. Its variance v_B serves as a very useful MLD filter tuning parameter. Increasing or decreasing its value raises or lowers, repectively, thresholds in our detectors (e.g., see Figures 4-9 and 4-10). This is illustrated in the following example.

Example 6-12. A one sec segment of stacked data consisting of 24 traces, obtained for a water gun source (courtesy of Unocal, Brea, California) is depicted in Figure 6-45. The datum point is sea level. Each trace is sampled at a 2 msec rate. Normal moveout correction and stacking have been applied with a 200% stretch mute.

Results of a sensitivity study to see the effects of v_B values on estimated wavelets and reflectivity are depicted in Figures 6-46 and 6-47. All results were obtained by processing the bottom trace of Figure 6-45 using a two phase Block Component Method. Estimated wavelets, in Figure 6-46, are for backscatter variances ranging from 0.1×10^{-4} to 0.35×10^{-3}. Observe that wavelet estimates do not depend on the backscatter variance, i.e., they are quite robust with respect to this parameter. Estimated reflectivity sequences, in Figure 6-47, have from 1 to 88 spikes depending upon v_B. See Table 6-2 for a complete summary of these results. Observe that by increasing v_B we decrease the number of detected spikes. It appears, therefore, that MLD can give a wide range of deconvolution results, ranging from MED, when v_B is large, to maximum entropy deconvolution, when v_B is smaller.

The data in Figure 6-45 was processed for v_B equal to 0.15×10^{-3} and 0.8×10^{-4}. The results are depicted in Figures 6-48 and 6-49, respectively. From Figure 6-48 we see that the estimated reflectivity sequences show reasonably good continuity, especially in view of a rather poor signal level at depth.

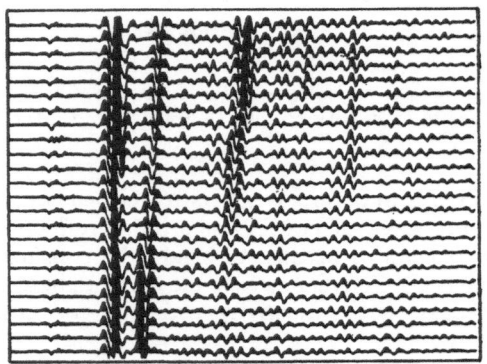

Figure 6-45. A one sec segment of stacked data.

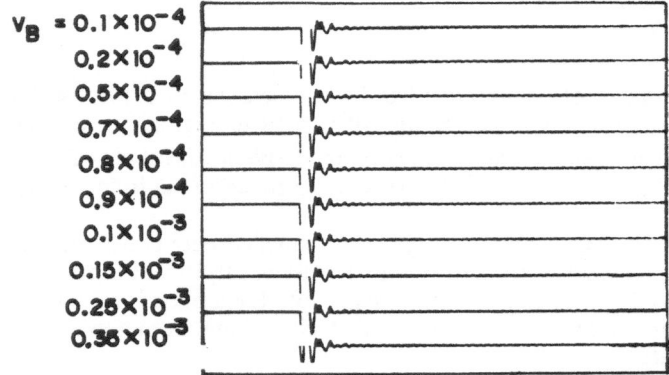

Figure 6-46. Estimated wavelets for 10 values of v_B, obtained by processing the bottom trace of Fig. 6-45.

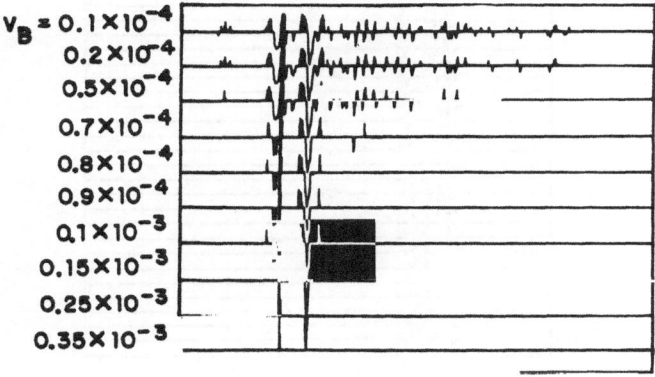

Figure 6-47. Estimated reflectivity sequences for 10 levels of backscatter, obtained by processing the bottom trace of Fig. 6-45.

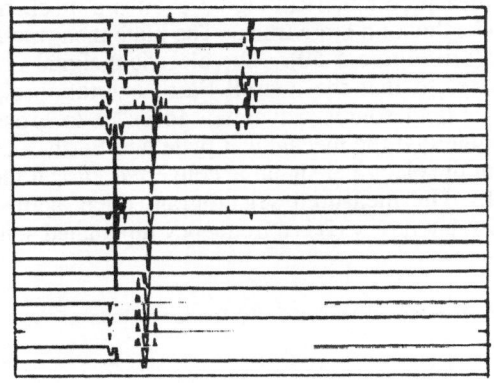

Figure 6-48. Processed input section for $v_B = 0.15 \times 10^{-3}$.

Comparing the estimated reflectivity sequences in Figures 6-48 and 6-49, we see that v_B can indeed be used to adjust the quality of the estimated reflectivity.

6.7 Noncausal Channel Models

All of the discussions in this book are applicable to a causal wavelet. Hsueh and Mendel (1985) have extended this book's results to noncausal wavelets, a

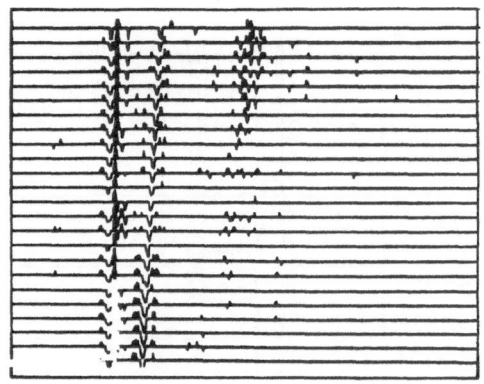

Figure 6-49. Processed input section for $v_B = 0.8 \times 10^{-4}$.

situation which occurs in the seismic area, for example, when a land vibrator [i.e., Vibroseis^{TM} (a trademark of the Continental Oil Co.)] is used.

In Vibroseis, a measured electrical signal excites a mechanical plate that is mounted on the underside of a large truck. The plate then provides a vibrational input to the Earth (see Figure 6-50). The electrical signal is swept through a range of frequencies, so that the overall Earth excitation signal looks quite different from other land source wavelets. In order to grossly simplify this discussion, we assume that the mechanical plate and Earth-coupling effects can be ignored, i.e., that the electrical signal is converted into an identically looking mechanical signal. The measured seismogram once again is given by the convolutional model, but now w(i) is a frequency modulated sinusoidal wavelet.

Table 6-2. Number of detected spikes for different values of v_B.

v_B	Detected Spikes	v_B	Detected Spikes
0.1×10^{-4}	88	0.9×10^{-4}	13
0.2×10^{-4}	83	0.1×10^{-3}	14
0.5×10^{-4}	54	0.15×10^{-3}	5
0.7×10^{-4}	19	0.25×10^{-3}	4
0.8×10^{-4}	14	0.35×10^{-3}	1

A customary step in Vibroseis processing is to preprocess the measurements by correlating them with the measured electrical signal w(i). Using z-transforms to express this operation, we obtain

$$Z_1(z) = Z(z)W(1/z) = W(z)W(1/z)U(z) + W(1/z)N(z) \qquad (6\text{-}6)$$

or

$$z_1(k) = h(k)*u(k) + n_1(k) \qquad (6\text{-}7)$$

where

$$h(k) = w(k)*w(-k) \qquad (6\text{-}8)$$

and we have defined $n_1(k) = w(-k)*n(k)$. Equation (6-7) is often the starting point for Vibroseis deconvolution (e.g., Lines and Clayton, 1977). Observe that the "wavelet" $h(k)$ is noncausal. In this case it is a zero-phase autocorrelation function. In this section we will refer to $z_1(k)$ and $n_1(k)$ as $z(k)$ and $n(k)$, respectively, for notational simplicity.

Noncausal source wavelets occur in other fields as well, such as absorption spectroscopy (Blass and Halsey, 1981), astronomical data analysis (Scargle, 1981), and absorption filters (Marschall, 1977).

Minimum-variance and maximum-likelihood deconvolution algorithms are developed in Hsueh and Mendel (1985) for either symmetrical or nonsymmetrical time-invariant source wavelets that are excited by either stationary or nonstationary white noise inputs. Batch minimum-variance deconvolution algorithms for a noncausal wavelet turn out to be structurally the same as those for a causal wavelet. Additionally, maximum-likelihood deconvolution algorithms for a noncausal wavelet, which involve event detection and amplitude restoration, are essentially the same as those for a causal wavelet. This is because the vector measurement equation in the noncausal wavelet case is still given by Equation (2-15), the only difference being that in the case of a causal wavelet the wavelet matrix $\mathbf{W}$ is given by the lower triangular matrix

$$\mathbf{W} = \begin{pmatrix} w(0) & 0 & \cdots & 0 \\ w(1) & w(0) & \cdots & 0 \\ \cdot & \cdot & \cdot & \cdot \\ \cdot & \cdot & \cdot & \cdot \\ \cdot & \cdot & \cdot & \cdot \\ w(N-1) & w(N-2) & \cdots & w(0) \end{pmatrix}$$

whereas in the case of a noncausal wavelet $\mathbf{W}$ is given by the full matrix

$$\mathbf{W} = \begin{pmatrix} h(0) & h(-1) & h(-2) & \cdots & h(-N+1) \\ h(1) & h(0) & h(-1) & \cdots & h(-N+2) \\ \cdot & \cdot & \cdot & \cdot & \cdot \\ \cdot & \cdot & \cdot & \cdot & \cdot \\ \cdot & \cdot & \cdot & \cdot & \cdot \\ h(N-1) & h(N-2) & h(N-3) & \cdots & h(0) \end{pmatrix}$$

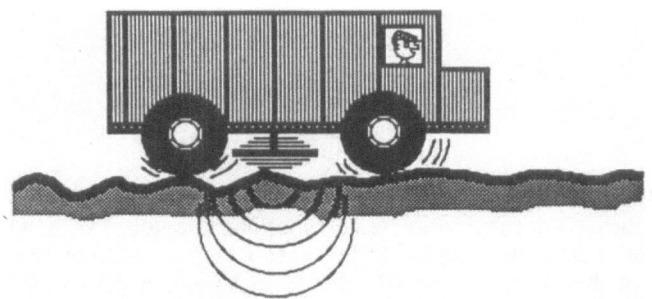

Figure 6-50. Vibroseis field experiment.

In both cases some of the elements in matrix W will equal zero because both w(i) and h(i) are truncated wavelets. Additionally, all of the results presented in Chapters 4 and 5 are derived in Chapter 7 just using the vector measurement equation (2-15) and its associated quantities. The exact structure of matrix W is never used in those derivations.

In Chapter 9 we discuss different computer implementations of the Chapter 7 batch formulas. A *recursive implementation* makes use of the structure of the W matrix; hence, a recursive implementation will have different looking formulas for the causal and noncausal wavelets. This is especially true for the MVD formulas   (see Hsueh and Mendel, 1985, for a comparison of these formulas).

The following examples are taken from Hsueh and Mendel (1985).

Example 6-13. Using a pseudo-random number generator, we generated the Bernoulli-Gaussian sequence, u(k), depicted in Figure 6-51. For this sequence $\lambda = 0.05$, N = 300 and there are 19 spikes. This signal was convolved with the symmetrical noncausal IR depicted in Figure 6-52 to which white Guassian noise was added [signal-to-noise ratio (SNR) is 10] to produce the synthetic data that is depicted in Figure 6-53.

Threshold and SMLR detectors were applied to the Figure 6-53 data. Figure 6-54 depicts the MVD estimates of u(k), which are used in the threshold detector to form $[u^{MV}(k \mid N)]^2$. The threshold detected estimates of u(k) are depicted in Figure 6-55.   Observe that there are 11 missed events, two false alarms, and one improperly located event at k = 139. Observe, also, that the amplitudes for the false alarms are quite small.

The threshold detected sequence was used to initialize the SMLR detector, which then ran for 14 iterations before it converged to the estimates depicted in Figure 6-56. Observe that there are two missed events and one improperly located event at k = 255, but there are no false alarms. Comparing Figures 6-55 and 6-56, we see that there is a large improvement in performance for the SMLR detector.

Comparing Figures 6-54 and 6-56, we see also that the MLD estimates of u(k) are much better than the MVD estimates. This is because, in MLD we make use of more information about u(k) than in MVD, i.e., in MLD we model u(k) as a Bernoulli-Gaussian sequence whereas in MVD we only use second-order statistics of u(k) (i.e., $v_u = \lambda v_r$).

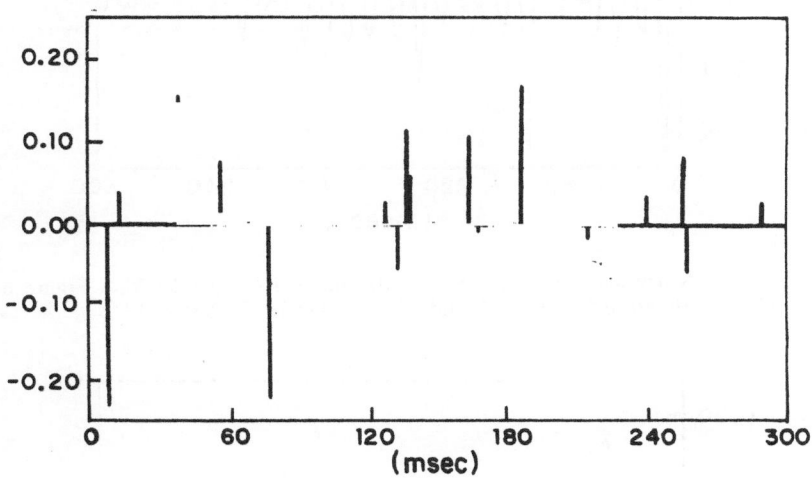

Figure 6-51.   True reflectivity sequence, u(k). (Hsueh and Mendel, 1985. © 1985 IEEE.)

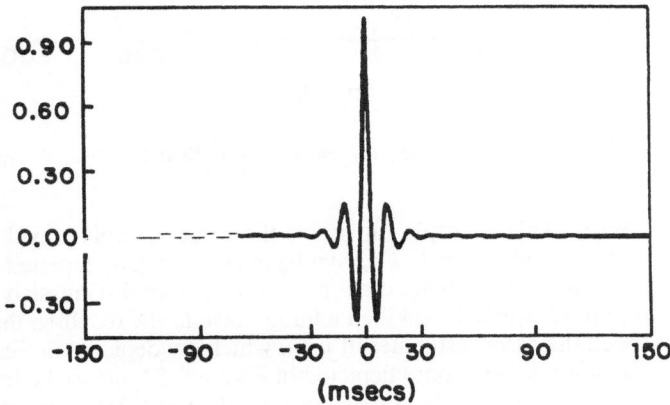

Figure 6-52.   Two-sided IR that is symmetric about the origin.   (Hsueh and Mendel, 1985.   © 1985 IEEE.)

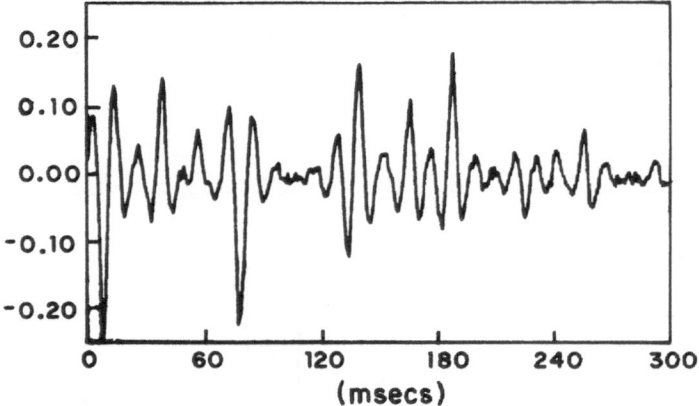

Figure 6-53. Noisy seismogram (SNR = 10) obtained using u(k) from Figure 6-51 and h(k) from Figure 6-52. (Hsueh and Mendel, 1985. © 1985 IEEE.)

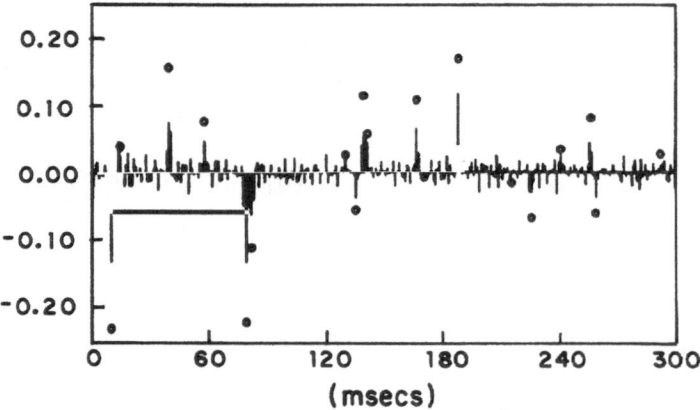

Figure 6-54. MVD estimates of u(k) for the Figure 6-53 data. (Hsueh and Mendel, 1985. © 1985 IEEE.)

Example 6-14. This example illustrates the effect of noise level on MLD results for a noncausal wavelet. Our synthetic seismogram, depicted in Figure 6-57, was obtained by convolving the Figure 6-51 Bernoulli-Gaussian sequence with the Figure 6-52 wavelet and then adding noise to the result so that SNR = 3. As we expect, the MVD estimates of u(k), which are depicted in Figure 6-58, are poorer in quality then the ones depicted in Figure 6-54 due to the lower SNR in the present case. The threshold detector only detected three true events (see Figure 6-59), and the SMLR detector took 20 iterations to converge to the results depicted in Figure 6-60. Observe that there are 10 detected events, one false alarm and 9 missed events; however, large events are detected. Clearly,

MLD performanced is enhanced as SNR increases. This is the same conclusion that was obtained for a causal wavelet.

Example 6-15. This example illustrates the effect of wavelet bandwidth on   MLD results. Our wavelet is the one depicted in Figure 6-61. Observe that it is narrower than the wavelet depicted in Figure 6-52; hence, it is of larger bandwidth than that wavelet.   Figure 6-62 depicts the   noisy

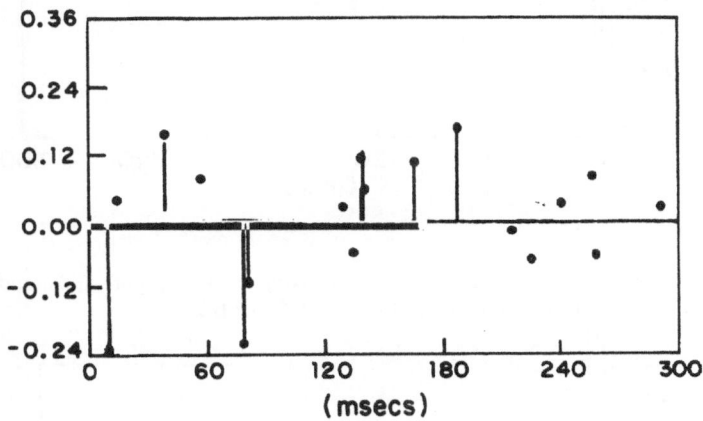

Figure 6-55. Threshold detected estimates of u(k) based on Figure 6-54 estimates. (Hsueh and Mendel, 1985. © 1985 IEEE.)

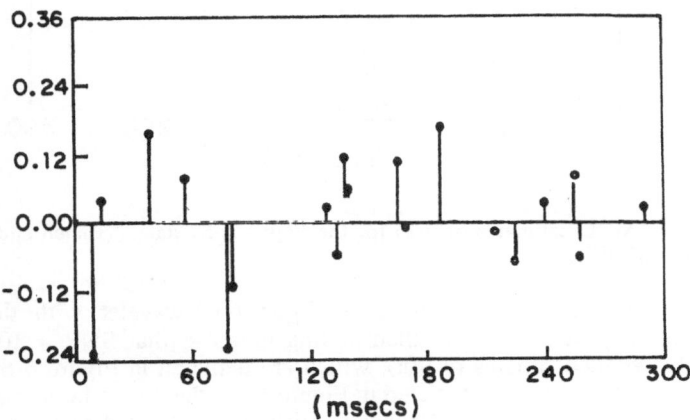

Figure 6-56. SMLR detected estimates of u(k) using Figure 6-55's results as an initial sequence. (Hsueh and Mendel,1985. © 1985 IEEE.)

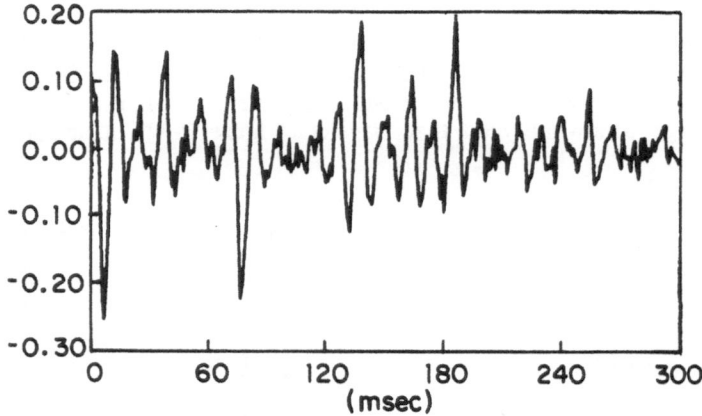

Figure 6-57. Noisy seismogram (SNR = 3) using u(k) in Figure 6-51 and h(k) in Figure 6-52. (Hsueh and Mendel, 1985. © 1985 IEEE.)

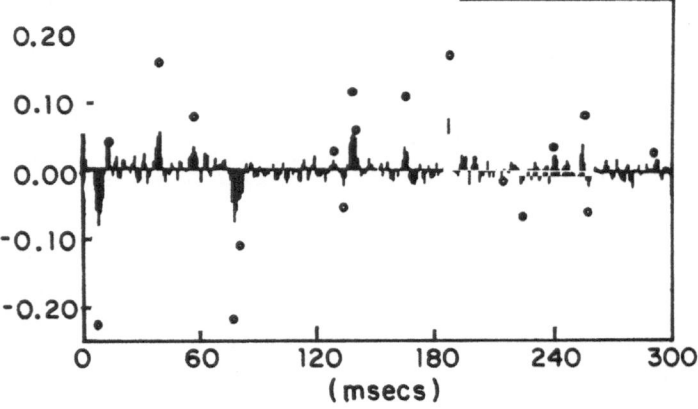

Figure 6-58. MVD estimates of u(k) for the Figure 6-57 data. (Hsueh and Mendel, 1985. © 1985 IEEE.)

seismogram, obtained by convolving the Figure 6-61 wavelet with the Figure 6-51 reflectivity sequence and then adding noise so that SNR = 10. As we expect, the MVD estimates of u(k), which are depicted in Figure 6-63, are of better quality than the ones depicted in Figure 6-54, due to the larger bandwidth of the Figure 6-62 wavelet. Consequently the threshold detector estimates, depicted in Figure 6-64, are also better than those depicted in Figure 6-55. In this case the SMLR detected estimates of u(k), depicted in Figure 6-65, were obtained in only three iterations (versus 14 iterations for the Figure 6-56 results) and are of excellent quality. Clearly, MLD performance is enhanced as wavelet bandwidth increases, just as in the case of a causal wavelet.☐

In reality, the seismic source signal that is transmitted into the earth is not the electrical signal that is applied to the mechanical plate. The electrical signal is distorted not only by the dynamics of the plate but also by effects of the unconsolidated earth upon which the rapping takes place. This distortion cannot be measured. Because it changes the electrical signal, it would seem necessary to estimate it if we are to achieve high-resolution Vibroseis deconvolution.

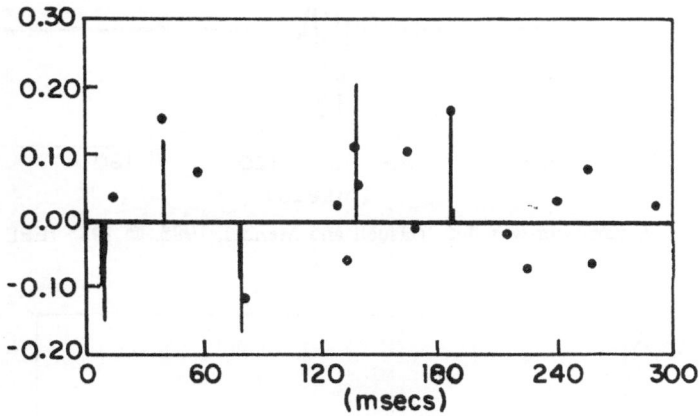

Figure 6-59. Threshold detected estimates of u(k), based on the MVD results in Figure 6-58. (Hsueh and Mendel, 1985. © 1985 IEEE.)

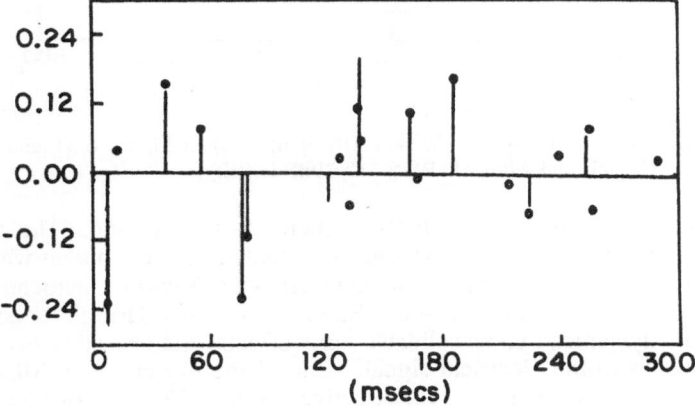

Figure 6-60. SMLR detected estimates of u(k) using Figure 6-59's results as an initial sequence. (Hsueh and Mendel,1985. © 1985 IEEE.)

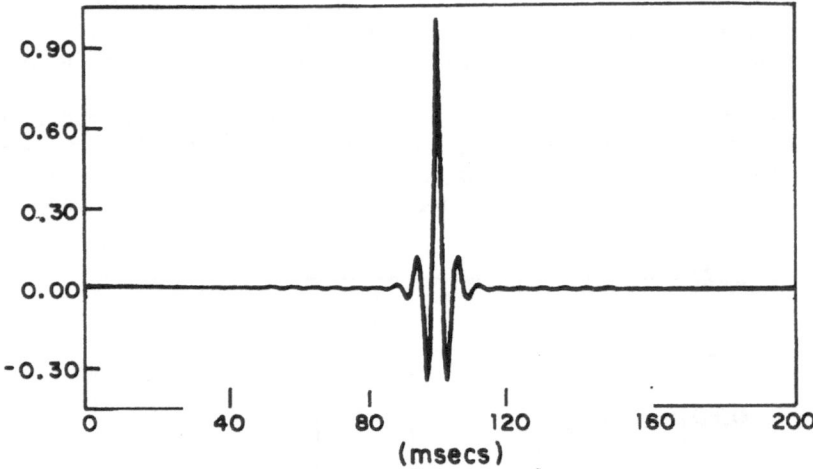

Figure 6-61. A narrower wavelet. (Hsueh and Mendel, 1985. © 1985 IEEE.)

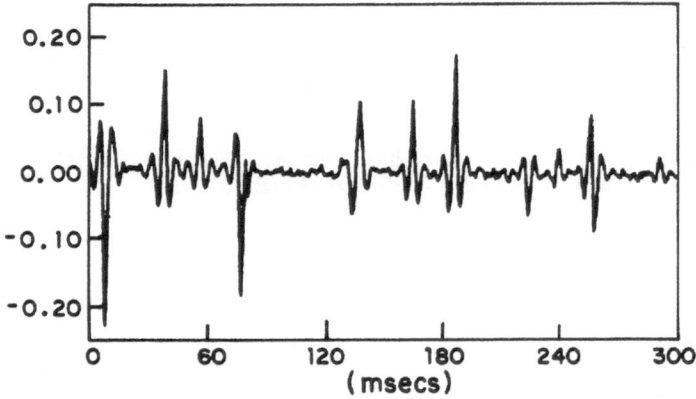

Figure 6-62. Noisy seismogram (SNR = 10) using u(k) in Figure 6-51 and h(k) in Figure 6-61. (Hsueh and Mendel, 1985. © 1985 IEEE.)

The situation is now the one that is depicted in Figure 6-66. The starting point for MLD is again the convolutional model; but, it is one in which the effective wavelet is the cascade of the noncausal ideal Vibroseis autocorrelation function and a filter which we call an "Earth Filter." This Earth Filter accounts for the plate dynamics , unconsolidated earth effects, and any other distortions that may occur to the electrical signal. One of the objectives in MLD is to estimate the unknown portion of the effective wavelet. The next two examples, which are taken from Hsueh (1983), illustrate some aspects of such partial wavelet estimation.

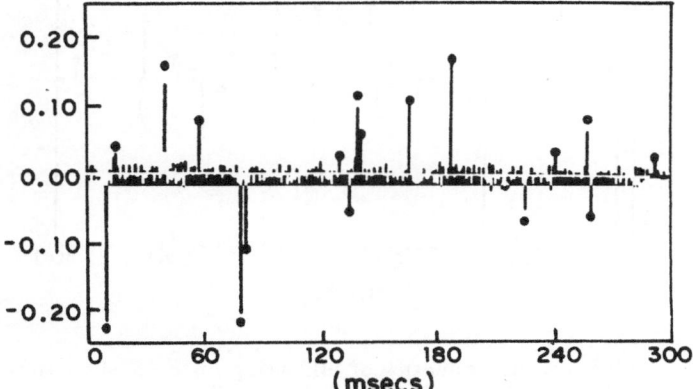

Figure 6-63. MVD estimates of u(k) for the Figure 6-62 data. (Hsueh and Mendel, 1985. © 1985 IEEE.)

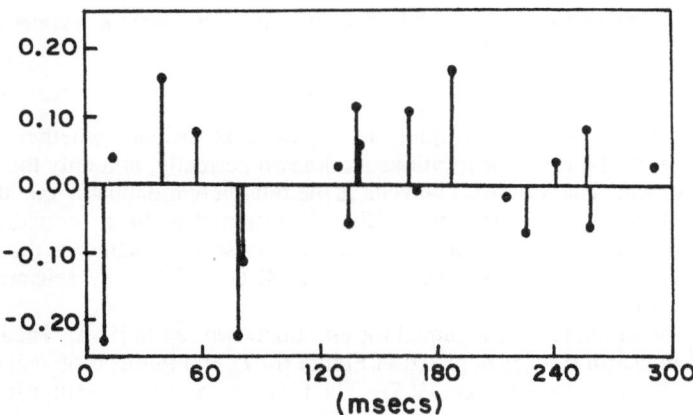

Figure 6-64. Threshold detected estimates of u(k), based on the results in Figure 6-63. (Hsueh and Mendel, 1985. © 1985 IEEE.)

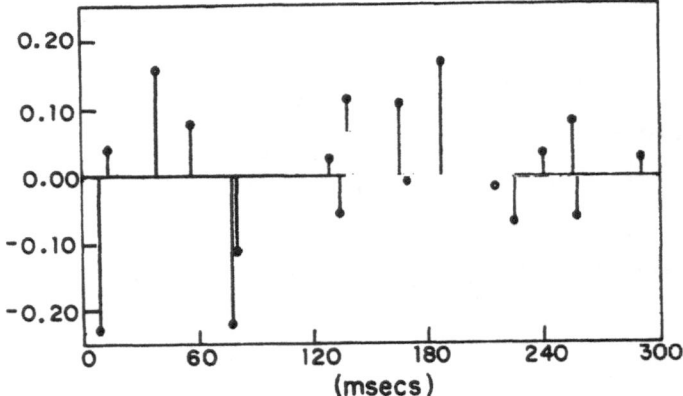

Figure 6-65. SMLR detected estimates of u(k) using Figure 6-64's results as an initial sequence. (Hsueh and Mendel, 1985. © 1985 IEEE.)

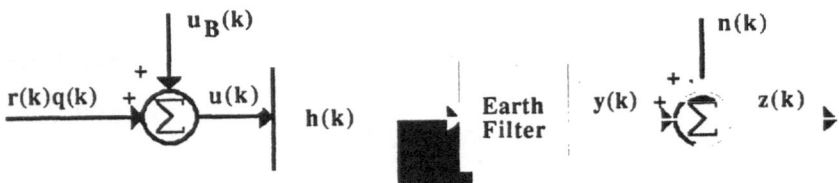

Figure 6-66. Modified convolutional model, in which the effective wavelet is the cascade of h(k) and the Earth Filter.

Example 6-16. In this example, the objective is to learn whether MLD works well when the reflector locations are known perfectly, and only the earth filter is unknown. The Vibroseis wavelet is the relatively broadband one that is depicted in Figure 6-52. The earth filter is assumed to be a second-order nonminimum phase ARMA, whose impulse response is depicted as the solid curve in Figure 6-67. The starting point for MLD is the noisy seismogram (SNR = 3) in Figure 6-68.

A second-order model was assumed for the "unknown" earth filter. The initial estimate of the earth filter is depicted in Figure 6-67. In Figures 6-69 and 6-70, we observe the final results from MLD. The final estimated earth filter is quite close to the actual earth filter, and the estimated reflectivity sequence is very close to the actual reflectivity sequence.

We conclude that, when reflector locations are known ahead of time and the Vibroseis wavelet is broadband, MLD works very well.

Hsueh has also shown that, if the Vibroseis wavelet is not broadband, then good results will not be obtained; hence, in Vibroseis deconvolution it is important to design the electrical sweep signal so that the Vibroseis wavelet is as broadband as possible.

Example 6-17. In this example, the primary objective is to learn whether MLD works well when the reflector locations are unknown, and the earth filter is also unknown. A secondary objective is to compare MLD with predictive deconvolution.

The Vibroseis wavelet is the broadband one depicted in Figure 6-52. The earth filter is now assumed to be the fourth-order nonminimum phase broadband wavelet that we have used in earlier examples. It is depicted in Figure 6-71, along with its initial estimate. The starting point for MLD is the noisy seismogram (SNR = 30) in Figure 6-72. The final MLD results are depicted in

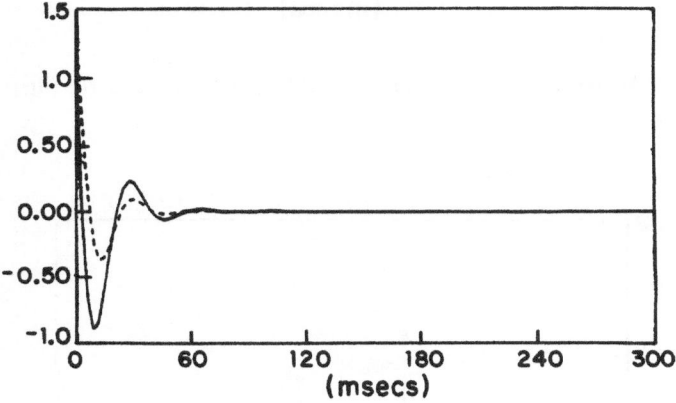

Figure 6-67. Second-order earth filter and its initial estimate (Hsueh, 1983).

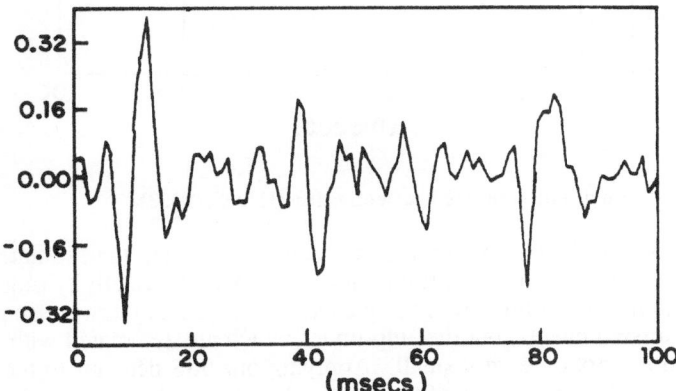

Figure 6-68. Noisy seismogram (SNR = 3) (Hsueh, 1983).

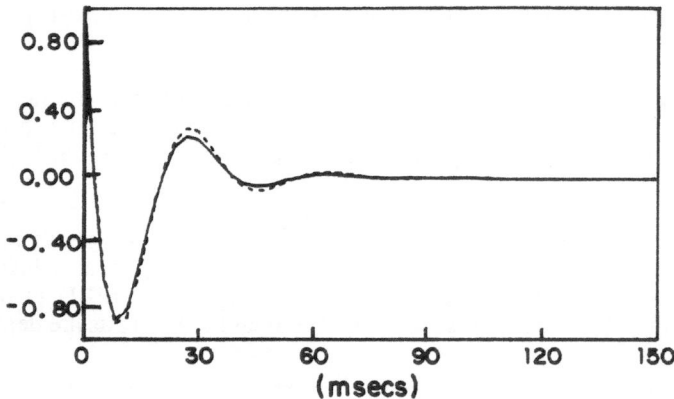

Figure 6-69. Actual (solid curve) and estimated (dashed curve) earth filters (Hsueh, 1983).

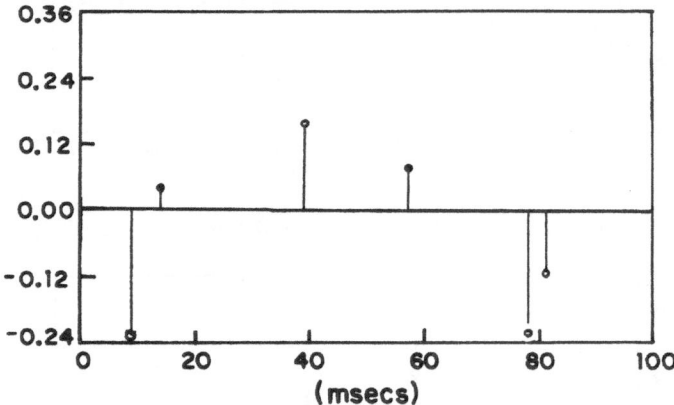

Figure 6-70. Actual and estimated reflectivities (Hsueh, 1983).

Figures 6-73 and 6-74. As in the previous example, the estimated earth filter converged to the true earth filter, and the estimated reflectivity sequence is quite close to the actual reflectivity sequence. Observe that there are some false alarms and missed events; but the amplitudes which are associated with both of these detection errors are very small. Applying our SSS detector to the Figure 6-74 results would cause many of the small false alarms that occur right next to a true detected event to merge with the true event.

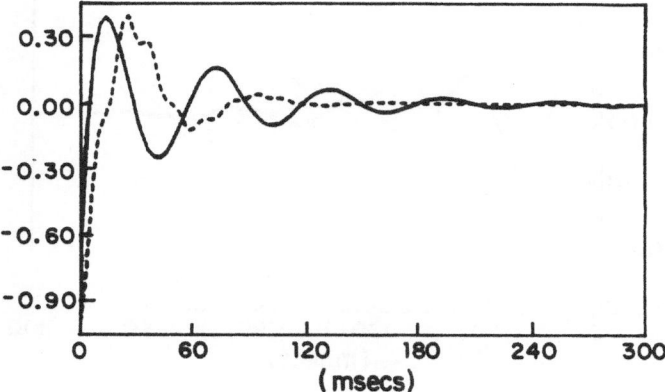

Figure 6-71. Fourth-order earth filter (solid curve) and its initial estimate (dashed curve) (Hsueh, 1983).

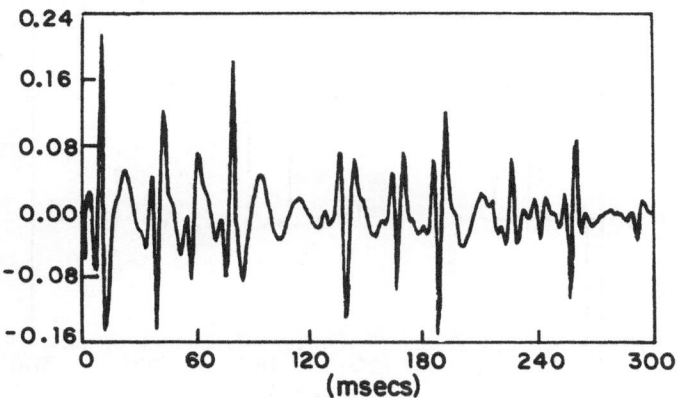

Figure 6-72. Noisy seismogram (SNR = 30) (Hsueh, 1983).

Results from predictive deconvolution (e.g., Robinson and Treitel, 1980) and MLD are compared in Figures 6-75 and 6-76. The MLD result was obtained by convolving the zero-phase Vibroseis wavelet with the final estimated reflectivity sequence. The results in Figure 6-76 appear to be much sharper than those in Figure 6-75. The predictive deconvolution results could, however, be used to provide a good initial event sequence that would expedite MLD. This would be accomplished by applying a threshold detector to the Figure 6-75 results.

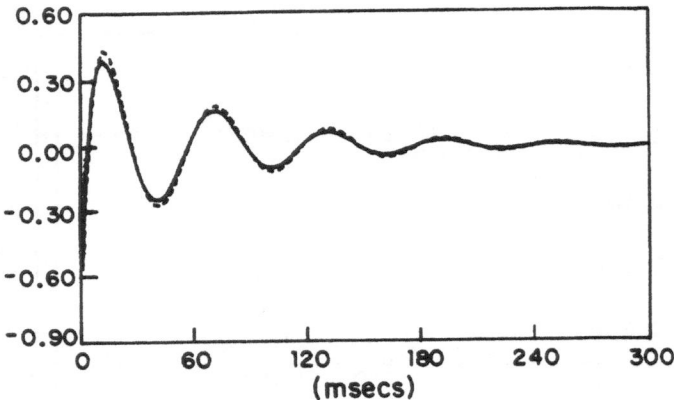

Figure 6-73. Actual (solid curve) and estimated (dashed curve) earth filters (Hsueh, 1983).

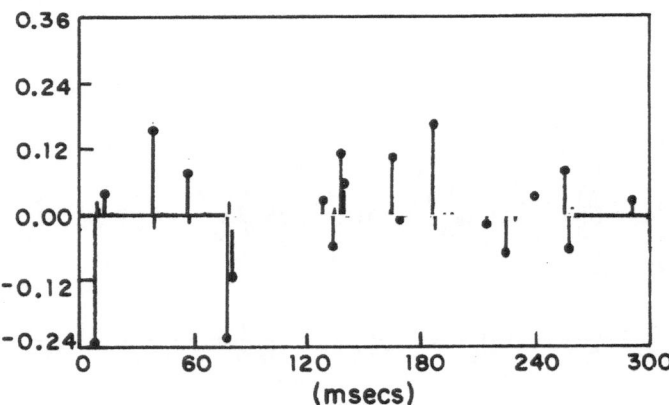

Figure 6-74. Actual and estimated reflectivity sequences (Hsueh, 1983).

6.8 Summary

The examples in this chapter have illustrated many aspects of the theory and properties of MLD. If we accept the original modeling assumption that the reflectivity sequence is non-Gaussian, then better performance must be obtained when nonlinear deconvolution is used instead of linear deconvolution.

Maximum-likelihood deconvolution is a nonlinear deconvolution procedure that is not only applicable to causal wavelets, but, is also applicable to noncausal wavelets. High resolution is possible because of the nonlinear nature of this deconvolution procedure. There is one common theme that connects many of this chapter's examples; namely, performance of MLD is limited, just as it is in any other deconvolution method, by the bandwidth of the seismic wavelet and by signal-to-noise ratio. However, it is less limited than in linear deconvolution techniques.

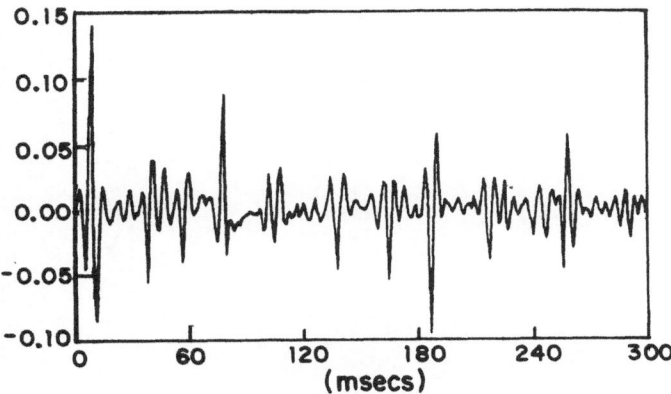

Figure 6-75. Predictive deconvolution results (Hsueh, 1983).

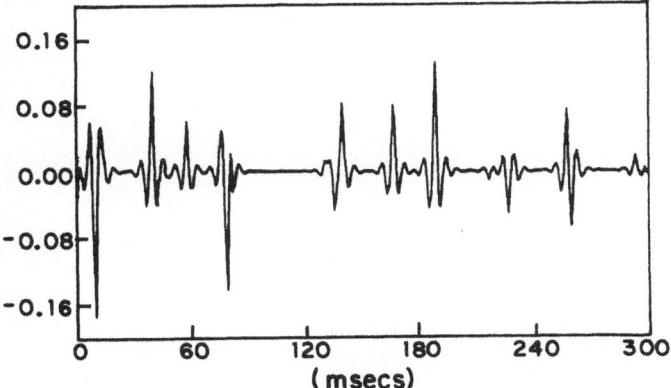

Figure 6-76. MLD result (Hsueh, 1983).

7
Mathematical Details
for Chapter 4

7.1 Introduction

This chapter is for the mathematically serious reader. In it we quantify many of the previous qualitative statements that were made in Chapter 4. For the convenience of the reader, we restate many of the Chapter 4 results, prior to their derivations.

7.2 Mathematical Fact

Here we provide a derivative-free proof of the fact that *the point* $(\hat{x}, \hat{y})$, *which maximizes* f(x,y) *can be found from the following four step procedure*:

1. *Set the partial derivative (or partial difference) of* f(x,y) *with respect to* x *equal to zero. For every* y, *call the solution* x, *that optimizes* f(x,y), h(y), *i.e.,* x = h(y). *If this solution cannot be obtained, then STOP.*

2. *Substitute* x = h(y) *into* f(x, y) *and call the resulting function* g(y), *i.e.,* g(y) = f[h(y), y].

3. *Find the point* $\hat{y}$ *that maximizes* g(y).

4. *Compute* $\hat{x}$ *as* $\hat{x} = h(\hat{y})$.

   Proof (Goutsias and Mendel,1986): Step 3 of the procedure tells us that we must find the point $\hat{y}$ that maximizes the function g(y). This means that we must choose $\hat{y}$ such that $g(y) \leqslant g(\hat{y})$. From Step 2 of the procedure, we know that g(y) = f[h(y), y]; hence,

$$f[h(y), y] \leqslant f[h(\hat{y}), \hat{y}].$$

From Step 4 of the procedure, we compute $\hat{x}$ as $\hat{x} = h(\hat{y})$. Substituting this expression into the preceding inequality, we see that

$$f[h(y), y] = f(x, y)\big|_{x = h(y)} \leqslant f(\hat{x}, \hat{y}).$$

From Step 1 of the procedure, we see that, for every fixed point y, the point x = h(y) is where the one-dimensional function $f(x, y)$, y = constant, attains its maximum value. This means that

$$f(x, y) \leqslant f(x, y)\big|_{x = h(y)},$$

for every y. Combining the last two inequalities, we finally obtain

$$f(x, y) \leqslant f(\hat{x}, \hat{y}),$$

which proves that by this four step procedure we are indeed able to find the point $(\hat{x}, \hat{y})$ that maximizes $f(x, y)$.

7.3 Separation Principle

Here we provide a proof of the very important and useful Separation Principle that: *instead of finding* $a^{ML}, b^{ML}, s^{ML}, q^{ML}, r^{ML},$ *and* $u_B{}^{ML}$ *directly by maximizing the loglikelihood function* $\mathcal{L}\{a, b, s, q, r, u_B \mid z\}$, *we can first find* $a^{ML}, b^{ML}, s^{ML},$ *and* q^{ML} *by maximizing a related function* $\mathcal{M}\{a, b, s, q \mid z\}$ *after which* r^{ML} *and* $u_B{}^{ML}$ *are easily computed as linear operations on the data z, i.e.,*

$$r^{ML} = A(a^{ML}, b^{ML}, s^{ML}, q^{ML})z \tag{7-1}$$

and

$$u_B{}^{ML} = B(a^{ML}, b^{ML}, s^{ML}, q^{ML})z. \tag{7-2}$$

More specifically, we shall show that (we have omitted the constant term)

$$\mathcal{M}\{a, b, s, q \mid z\} = -\frac{N}{2}\ln(v_n v_r v_B) - \frac{1}{2}z'\Omega^{-1}z$$
$$+ m(q)\ln(\lambda) + [N - m(q)]\ln(1 - \lambda) \tag{7-3}$$

where

$$\Omega = v_r \mathbf{WQW'} + v_B \mathbf{WW'} + v_n \mathbf{I} , \qquad (7\text{-}4)$$

and $m(q)$ is defined in Eq. (3-6). We shall also show, that

$$\mathbf{r}^{ML} = v_r{}^{ML} \mathbf{Q}^{ML} \mathbf{W'}^{ML} (\Omega^{ML})^{-1} \mathbf{z}, \qquad (7\text{-}5)$$

and

$$\mathbf{u_B}^{ML} = v_B{}^{ML} \mathbf{W'}^{ML} (\Omega^{ML})^{-1} \mathbf{z}. \qquad (7\text{-}6)$$

In Eqs. (7-5) and (7-6), Ω^{ML} is obtained from Eq. (7-4) by replacing v_r, $\mathbf{W}$, $\mathbf{Q}$, v_B, and v_n with their ML values.

Proof: As noted in Chapter 4, we prove the Separation Principle by using the Mathematical Fact in which we set $\mathbf{x}$ = elements of $\mathbf{r}$ and $\mathbf{u_B}$ and $\mathbf{y}$ = elements of $\mathbf{a}$, $\mathbf{b}$, $\mathbf{s}$, and $\mathbf{q}$. To begin, recall Eq. (3-10) for the loglikelihood function $\mathcal{L}\{\mathbf{a}, \mathbf{b}, \mathbf{s}, \mathbf{q}, \mathbf{r}, \mathbf{u_B} \mid \mathbf{z}\}$:

$$
\begin{aligned}
\mathcal{L}\{\mathbf{a}, \mathbf{b}, \mathbf{s}, \mathbf{q}, \mathbf{r}, \mathbf{u_B} \mid \mathbf{z}\} = & -\tfrac{N}{2} \ln(v_n v_r v_B) - \mathbf{r'r}/2v_r \\
& - (\mathbf{z} - \mathbf{WQr} - \mathbf{Wu_B})'(\mathbf{z} - \mathbf{WQr} - \mathbf{Wu_B})/2v_n \\
& - \mathbf{u_B' u_B}/2v_B + m(q)\ln(\lambda) \\
& + [N - m(q)]\ln(1 - \lambda).
\end{aligned}
\qquad (7\text{-}7)
$$

The gradients of $\mathcal{L}\{\ \}$ with respect to $\mathbf{r}$ and $\mathbf{u_B}$ are:

$$\partial\mathcal{L}\{\mathbf{a}, \mathbf{b}, \mathbf{s}, \mathbf{q}, \mathbf{r}, \mathbf{u_B} \mid \mathbf{z}\}/\partial\mathbf{r} = -\mathbf{r}/v_r + (\mathbf{WQ})'(\mathbf{z} - \mathbf{WQr} - \mathbf{Wu_B})/v_n$$

and

$$\partial\mathcal{L}\{\mathbf{a}, \mathbf{b}, \mathbf{s}, \mathbf{q}, \mathbf{r}, \mathbf{u_B} \mid \mathbf{z}\}/\partial\mathbf{u_B} = -\mathbf{u_B}/v_B + \mathbf{W'}(\mathbf{z} - \mathbf{WQr} - \mathbf{Wu_B})/v_n .$$

Setting these gradients equal to zero, and calling the resulting values of $\mathbf{r}$ and $\mathbf{u_B}$ $\mathbf{r}^*$ and $\mathbf{u_B}^*$, respectively, we obtain the following two simultaneous equations for $\mathbf{r}^*$ and $\mathbf{u_B}^*$:

$$-\mathbf{r}^*/v_r + (\mathbf{WQ})'(\mathbf{z} - \mathbf{WQr}^* - \mathbf{Wu_B}^*)/v_n = 0 \qquad (7\text{-}8)$$

and

$$-\mathbf{u_B}^*/v_B + \mathbf{W'}(\mathbf{z} - \mathbf{WQr}^* - \mathbf{Wu_B}^*)/v_n = 0. \qquad (7\text{-}9)$$

These equations can be written more succinctly, as

$$(v_n \mathbf{I} + v_r \mathbf{QW'WQ})\mathbf{r}^* + v_r \mathbf{QW'W u_B}^* = v_r \mathbf{QW'z} \qquad (7\text{-}10)$$

and

$$v_B \mathbf{W'WQ r}^* + (v_n \mathbf{I} + v_B \mathbf{W'W})\mathbf{u_B}^* = v_B \mathbf{W'z}. \qquad (7\text{-}11)$$

Multiply Eq. (7-11) by $Q v_r / v_B$ to see that

$$v_r Q W'WQr^* + [(v_r v_n / v_B)Q + v_r Q W'W]u_B^* = v_r Q W'z. \qquad (7\text{-}12)$$

Now, subtract Eq. (7-12) from Eq. (7-10) to obtain the following simple relationship between r^* and u_B^*:

$$(v_r v_n / v_B)Q u_B^* - v_n r^* = 0,$$

from which we find, that

$$r^* = (v_r / v_B)Q u_B^* . \qquad (7\text{-}13)$$

Next, substitute Eq. (7-13) into Eq. (7-11) [noting that $Q^2 = Q$], to see that

$$(v_n I + v_B W'W + v_r W'WQ)u_B^* = v_B W'z,$$

from which it follows, that

$$u_B^* = v_B(v_n I + v_B W'W + v_r W'WQ)^{-1}W'z. \qquad (7\text{-}14)$$

Applying Eq. (7-13) to this result, we find that

$$r^* = v_r Q(v_n I + v_B W'W + v_r W'WQ)^{-1}W'z. \qquad (7\text{-}15)$$

Observe from these results that r^* and u_B^* are indeed linear transformations of the data z.

At this point we have shown that it is indeed possible to express r^* and u_B^* as functions of $a, b, s,$ and q. Next, we must substitute r^* and u_B^* into the loglikelihood function $\mathcal{L}\{a, b, s, q, r, u_B \mid z\}$ in Eq. (7-7), in order to obtain an explicit formula for the related function $\mathcal{M}\{a, b, s, q, \mid z\}$. Before doing this, it turns out that there is another way to express r^* and u_B^* that we have found to be more useful than Eqs. (7-14) and (7-15). It is in terms of the covariance matrix of the data vector z, that is conditioned on knowing the event vector q, or equivalently, matrix Q. Let

$$\Omega = E\{zz' \mid Q\} ; \qquad (7\text{-}16)$$

then, substituting Eq. (2-19) for z into Eq. (7-16), we find

$$\Omega = E\{(WQr + Wu_B + n)(WQr + Wu_B + n)' \mid Q\}$$
$$\Omega = E\{(WQrr'QW' + WQru_B'W' + WQrn' + Wu_B r'QW' + Wu_B u_B'W'$$
$$+ Wu_B n' + nr'QW' + nu_B'W' + nn' \mid Q\}$$
$$\Omega = v_r WQW' + v_B WW' + v_n I. \qquad (7\text{-}17)$$

In obtaining Eq. (7-17) we used the facts that $\mathbf{r}$ and $\mathbf{u_B}$, $\mathbf{r}$ and $\mathbf{n}$, and $\mathbf{u_B}$ and $\mathbf{n}$ are uncorrelated and are zero mean [so that, for example, $E\{\mathbf{r}\mathbf{u_B}'\} = 0$], $Q^2 = Q$, $E\{\mathbf{r}\mathbf{r}'\} = v_r\mathbf{I}$, etc.

The following identity is true:

$$(v_n\mathbf{I} + v_B\mathbf{W'W} + v_r\mathbf{W'WQ})^{-1}\mathbf{W'} = \mathbf{W'\Omega^{-1}} . \tag{7-18}$$

This is most easily demonstrated by reexpressing Eq. (7-18) as

$$\mathbf{W'\Omega} = (v_n\mathbf{I} + v_B\mathbf{W'W} + v_r\mathbf{W'WQ})\mathbf{W'},$$

and substituting Eq. (7-17) for Ω into this equation, to show that the left-hand side does indeed equal the right-hand side; hence, we can use Eq. (7-18) to reexpress Eqs. (7-14) and (7-15), as

$$\mathbf{u_B}^* = v_B\mathbf{W'\Omega^{-1}z} \tag{7-19}$$

and

$$\mathbf{r}^* = v_r\mathbf{QW'\Omega^{-1}z}. \tag{7-20}$$

Now we are ready to substitute $\mathbf{u_B}^*$ and $\mathbf{r}^*$ into $\mathscr{L}\{\mathbf{a}, \mathbf{b}, \mathbf{s}, \mathbf{q}, \mathbf{r}, \mathbf{u_B} \mid \mathbf{z}\}$ in Eq. (7-7), in order to obtain an explicit formula for the related function $\mathscr{M}\{\mathbf{a}, \mathbf{b}, \mathbf{s}, \mathbf{q}, \mid \mathbf{z}\}$, i.e.,

$$\mathscr{M}\{\mathbf{a}, \mathbf{b}, \mathbf{s}, \mathbf{q}, \mid \mathbf{z}\} = \mathscr{L}\{\mathbf{a}, \mathbf{b}, \mathbf{s}, \mathbf{q}, \mathbf{r}^*, \mathbf{u_B}^* \mid \mathbf{z}\} . \tag{7-21}$$

We begin by reexpressing $\mathscr{L}\{\ \}$ as

$$\mathscr{L}\{\mathbf{a}, \mathbf{b}, \mathbf{s}, \mathbf{q}, \mathbf{r}^*, \mathbf{u_B}^* \mid \mathbf{z}\} = -\frac{N}{2}\ \ln(v_nv_rv_B) - T_1$$
$$+ m(\mathbf{q})\ln(\lambda) + [N - m(\mathbf{q})]\ln(1 - \lambda) , \tag{7-22}$$

where

$$T_1 = \mathbf{r}^{*\prime}\mathbf{r}^*/2v_r + (\mathbf{z} - \mathbf{WQr}^* - \mathbf{Wu_B}^*)'(\mathbf{z} - \mathbf{WQr}^* - \mathbf{Wu_B}^*)/2v_n$$
$$+ \mathbf{u_B}^{*\prime}\mathbf{u_B}^*/2v_B . \tag{7-23}$$

Expanding this equation, and grouping terms, it is easy to show that

$$T_1 = \mathbf{r}^{*\prime}\mathbf{r}^*/2v_r + \mathbf{z}'(\mathbf{z} - \mathbf{WQr}^* - \mathbf{Wu_B}^*)/2v_n - \mathbf{r}^{*\prime}(\mathbf{WQ})'(\mathbf{z} - \mathbf{WQr}^*$$
$$- \mathbf{Wu_B}^*)/2v_n - \mathbf{u_B}^{*\prime}\mathbf{W'}(\mathbf{z} - \mathbf{WQr}^* - \mathbf{Wu_B}^*)/2v_n + \mathbf{u_B}^{*\prime}\mathbf{u_B}^*/2v_B.$$

The first and third terms cancel, by virtue of Eq. (7-8), and the fourth and fifth terms cancel by virtue of Eq. (7-9); hence,

$$T_1 = \mathbf{z}'(\mathbf{z} - \mathbf{WQr}^* - \mathbf{Wu_B}^*)/2v_n . \tag{7-24}$$

We now expand the right-hand side of Eq. (7-24) to show that $T_1 = \frac{1}{2}z'\Omega^{-1}z$. Proceeding with the expansion, we see that

$$T_1 = z'z/2v_n - z'W(Qr^* + u_B^*)/2v_n ; \qquad (7\text{-}25)$$

but, from Eq. (7-13), we know that

$$Qr^* = (v_r/v_B)Qu_B^*;$$

consequently, Eq. (7-25) can be written as

$$T_1 = z'z/2v_n - z'W[(v_r/v_B)Q + I]u_B^*/2v_n . \qquad (7\text{-}26)$$

Substituting Eq. (7-19) for u_B^* into this equation, we find that

$$T_1 = z'z/2v_n - z'W[(v_r/v_B)Q + I]\, \tilde{v}_B W'\Omega^{-1}z/2v_n$$
$$T_1 = z'[I - v_r WQW'\Omega^{-1} - v_B WW'\Omega^{-1}]z/2v_n$$
$$T_1 = z'[\Omega - v_r WQW' - v_B WW']\Omega^{-1}z/2v_n . \qquad (7\text{-}27)$$

Substituting Eq. (7-17) for Ω into this last equation, many cancellations occur, and we obtain the desired result, that

$$T_1 = z'\Omega^{-1}z/2. \qquad (7\text{-}28)$$

Substituting this expression for T_1 into Eq. (7-22), and calling the result $\mathcal{M}\{a,b,s,q, \mid z\}$, we obtain the explicit formula for $\mathcal{M}\{a,b,s,q, \mid z\}$ that is given in Eq. (7-3). It is this function that is maximized to find a^{ML}, b^{ML}, s^{ML}, and q^{ML}, after which r^{ML} and u_B^{ML} are computed by substituting these values into Eqs. (7-20) and (7-19), respectively, i.e.,

$$r^{ML} = r^*[a = a^{ML}, b = b^{ML}, s = s^{ML}, q = q^{ML}],$$
$$r^{ML} = v_r^{ML}Q^{ML}W'^{ML}(\Omega^{ML})^{-1}z , \qquad (7\text{-}29)$$

and

$$u_B^{ML} = u_B^*[a = a^{ML}, b = b^{ML}, s = s^{ML}, q = q^{ML}] ,$$
$$u_B^{ML} = v_B^{ML}W'^{ML}(\Omega^{ML})^{-1}z , \qquad (7\text{-}30)$$

where

$$W^{ML} = W(a^{ML}, b^{ML}) \qquad (7\text{-}31)$$

and

$$\Omega^{ML} = v_r^{ML}W^{ML}Q^{ML}W'^{ML} + v_B^{ML}W^{ML}W'^{ML} + v_n^{ML}I. \qquad (7\text{-}32)$$

It is clear from these equations that the matrices $\mathbf{A}(\mathbf{a}^{ML}, \mathbf{b}^{ML}, \mathbf{s}^{ML}, \mathbf{q}^{ML})$ and $\mathbf{B}(\mathbf{a}^{ML}, \mathbf{b}^{ML}, \mathbf{s}^{ML}, \mathbf{q}^{ML})$, which appear in the statement of the Separation Principle, are

$$\mathbf{A}(\mathbf{a}^{ML}, \mathbf{b}^{ML}, \mathbf{s}^{ML}, \mathbf{q}^{ML}) = v_r^{ML} \mathbf{Q}^{ML} \mathbf{W}'^{ML} (\Omega^{ML})^{-1}$$

and

$$\mathbf{B}(\mathbf{a}^{ML}, \mathbf{b}^{ML}, \mathbf{s}^{ML}, \mathbf{q}^{ML}) = v_B^{ML} \mathbf{W}'^{ML} (\Omega^{ML})^{-1}$$

This completes the proof of the Separation Principle.☐

It is easy to compute the maximum-likelihood value of the entire input, $\mathbf{u}$, namely, $\mathbf{u}^{ML}$, from the Invariance Property of maximum likelihood and, $\mathbf{r}^{ML}$ and $\mathbf{u}_B^{ML}$, i.e.,

$$\mathbf{u}^{ML} = \mathbf{Q}^{ML} \mathbf{r}^{ML} + \mathbf{u}_B^{ML} = \mathbf{r}^{ML} + \mathbf{u}_B^{ML} \tag{7-33}$$

The fact that $\mathbf{Q}^{ML} \mathbf{r}^{ML} = \mathbf{r}^{ML}$ follows directly from Equation (7-29) and the fact that $(\mathbf{Q}^{ML})^2 = \mathbf{Q}^{ML}$. Note that we use our knowledge of $\mathbf{Q}^{ML}$ to compute both $\mathbf{r}^{ML}$ and $\mathbf{u}_B^{ML}$; $\mathbf{Q}^{ML}$ does not disappear in the calculation of $\mathbf{u}^{ML}$.

Note, also, that it is not necessary to compute both $\mathbf{r}^{ML}$ and $\mathbf{u}_B^{ML}$; for,

$$\mathbf{r}^{ML} = (v_r^{ML}/v_B^{ML}) \mathbf{Q}^{ML} \mathbf{u}_B^{ML} \tag{7-33'}$$

7.4 Minimum Variance Deconvolution

Values of $u(k)$ that are a function of the data are known as *estimates* of $u(k)$. They may be linear or nonlinear functions of the data. Because a threshold detector begins with *linear minimum-variance estimates* of $u(k)$, we precede our discussions about it with a derivation of the minimum-variance estimator of $u(k)$, which we denote $u^{MV}(k \mid N)$. The conditioning denotes the fact that all N measurements, $z(1), z(2), \ldots, z(N)$, are used to obtain estimates of each $u(k)$, $k = 1, 2, \ldots, N$. Note that the term *estimator* denotes a filter structure. When actual measurements are input to an estimator its outputs are *estimates*.

A *linear minimum-variance estimator* of $u(k)$ is also known as a *Minimum-Variance Deconvolution (MVD) filter*. It is important to realize that $\mathbf{a}$, $\mathbf{b}$, and $\mathbf{s}$ are assumed to be known during the design and implementation of the MVD filter.

Instead of determining $u^{MV}(k \mid N)$ separately for each value of $k = 1, 2, \ldots, N$, we shall first determine

$$\mathbf{u}^{MV}(N) = \text{col}[u^{MV}(1 \mid N), u^{MV}(2 \mid N), \ldots, u^{MV}(N \mid N)], \tag{7-34}$$

and then pull out a general expression for the k th element of $\mathbf{u}^{MV}(N)$, namely, $\mathbf{u}^{MV}(k \mid N)$. We shall do the same for the covariance matrix that describes the errors between $\mathbf{u}$ and $\mathbf{u}^{MV}(N)$.

We shall show that

$$\mathbf{u}^{MV}(N) = \mathbf{P}_u \mathbf{W}'(\mathbf{W}\mathbf{P}_u\mathbf{W}' + v_n\mathbf{I})^{-1}\mathbf{z} , \qquad (7\text{-}35)$$

$$\mathbf{u}^{MV}(k \mid N) = v_u(k)\mathbf{w}'_k(\mathbf{W}\mathbf{P}_u\mathbf{W}' + v_n\mathbf{I})^{-1}\mathbf{z} , \qquad (7\text{-}36)$$

$$\text{Covar}[\mathbf{u}_{ERR}(N)] = \mathbf{P}_u - \mathbf{P}_u\mathbf{W}'(\mathbf{W}\mathbf{P}_u\mathbf{W}' + v_n\mathbf{I})^{-1}\mathbf{W}\mathbf{P}_u , \qquad (7\text{-}37)$$

and

$$\text{Var}[\mathbf{u}_{ERR}(k \mid N)] = v_u(k) - v_u^2(k)\mathbf{w}'_k(\mathbf{W}\mathbf{P}_u\mathbf{W}' + v_n\mathbf{I})^{-1}\mathbf{w}_k , \qquad (7\text{-}38)$$

where $v_u(k)$, *the diagonal element of input covariance matrix* $\mathbf{P}_u$, *is the variance of* $u(k)$,

$$\mathbf{u}_{ERR}(N) = \mathbf{u} - \mathbf{u}^{MV}(N), \qquad (7\text{-}39)$$

and $\mathbf{w}_k$ *is the k th column of matrix* $\mathbf{W}$.

Proof: We begin by assuming that $\mathbf{u}^{MV}(N)$ is a linear function of $\mathbf{z}$, i.e.,

$$\mathbf{u}^{MV}(N) = \mathbf{K}_L(N)\mathbf{z} , \qquad (7\text{-}40)$$

where $\mathbf{K}_L(N)$ is an $N{\times}N$ gain matrix that is chosen to minimize the error variance between $\mathbf{u}$ and $\mathbf{u}^{MV}(N)$. This is yet another optimization problem. We shall choose $\mathbf{K}_L(N)$ to minimize the sum of the variances of the elements of $\mathbf{u}_{ERR}(N)$ (because variances are positive, minimizing the sum of the variances is equivalent to minimizing each of the individual variances). We let $\sum \text{Var}[\mathbf{u}_{ERR}(N)]$ denote the sum of the variances, i.e.,

$$\sum \text{Var}[\mathbf{u}_{ERR}(N)] = \mathbf{E}\{[\mathbf{u}_{ERR}(N) - \mathbf{E}\{\mathbf{u}_{ERR}(N)\}]'[\mathbf{u}_{ERR}(N) - \mathbf{E}\{\mathbf{u}_{ERR}(N)\}]\} .$$

Note that

$$\mathbf{E}\{\mathbf{u}_{ERR}(N)\} = \mathbf{E}\{\mathbf{u}\} - \mathbf{E}\{\mathbf{u}^{MV}(N)\} = -\mathbf{K}_L(N)\mathbf{E}\{\mathbf{z}\} = -\mathbf{K}_L(N)\mathbf{E}\{\mathbf{W}\mathbf{u} + \mathbf{n}\}$$
$$\mathbf{E}\{\mathbf{u}_{ERR}(N)\} = -\mathbf{K}_L(N)\mathbf{W}\mathbf{E}\{\mathbf{u}\} - \mathbf{K}_L(N)\mathbf{E}\{\mathbf{n}\} = 0 , \qquad (7\text{-}41)$$

because $\mathbf{u}$ and $\mathbf{n}$ are zero mean vectors; hence,

$$\sum \text{Var}[\mathbf{u}_{ERR}(N)] = \mathbf{E}\{\mathbf{u}_{ERR}'(N)\ \mathbf{u}_{ERR}(N)\} . \qquad (7\text{-}42)$$

Substituting $\mathbf{u}^{MV}(N)$ from Eq. (7-40) into this expression, we find that

$$\sum \text{Var}[u_{ERR}(N)] = E\{[u - K_L(N)z]'[u - K_L(N)z]\}$$
$$= \text{trace } E\{[u - K_L(N)z][u - K_L(N)z]'\}$$
$$= \text{trace } E\{uu' - uz'K_L'(N) - K_L(N)zu' + K_L(N)zz'K_L'(N),$$
$$\sum \text{Var}[u_{ERR}(N)] = \text{trace } E\{uu'\} + \text{trace } [K_L(N)E\{zz'\}K_L'(N)$$
$$- E\{uz'\}K_L(N) - K_L(N)E\{zu'\}] . \tag{7-43}$$

Let P_u, P_z, P_{uz}, and P_{zu} denote the covariance matrices of u and z, and the cross-covariance matrices between u and z and z and u, respectively. Note that $P_{zu} = [P_{uz}]'$. Equation (7-43) can be expressed in terms of these matrices, as

$$\sum \text{Var}[u_{ERR}(N)] = \text{trace } P_u + \text{trace } [K_L(N)P_z K_L'(N)$$
$$- P_{uz}K_L'(N) - K_L(N)P_{zu}]$$
$$= \text{trace } P_u + \text{trace}\{[K_L(N)P_z - P_{uz}]P_z^{-1}$$
$$[K_L(N)P_z - P_{uz}]' - P_{uz}P_z^{-1}P_{uz}'\} . \tag{7-44}$$

The error variance is minimized [with respect to $K_L(N)$] when the argument of the second trace function is minimized, which occurs when $K_L(N)P_z - P_{uz} = 0$, or

$$K_L(N) = P_{uz}P_z^{-1}. \tag{7-45}$$

Substituting Eq. (7-45) into Eq. (7-40), we obtain the following formula for $u^{MV}(N)$:

$$u^{MV}(N) = P_{uz}P_z^{-1}z . \tag{7-46}$$

We must now evaluate P_{uz} and P_z. Using the definition of a covariance function and the model for z, we find that

$$P_{uz} = E\{[u - E\{u\}][z - E\{z\}]'\} = E\{uz'\} = E\{u(Wu + n)'\}$$
$$= E\{uu'\}W' + E\{un'\} = P_u W' , \tag{7-47}$$

because u and n are statistically independent and are zero mean. Depending on what our input model is, P_u can have quite different values. We explore this below.

Continuing, we find that

$$P_z = E\{[z - E\{z\}][z - E\{z\}]'\} = E\{zz'\}$$
$$= E\{(Wu + n)(Wu + n)'\} = WE\{uu'\}W' + WE\{un'\}$$
$$+ E\{nu'\}W' + E\{nn'\} = WP_u W' + v_n I . \tag{7-48}$$

Substituting Eqs. (7-47) and (7-48) into Eq. (7-46), we obtain the final result, that

$$u^{MV}(N) = P_u W'(WP_u W' + v_n I)^{-1}z . \tag{7-49}$$

To obtain the k th component of $\mathbf{u}^{MV}(N)$, namely $u^{MV}(k \mid N)$, note that matrix $\mathbf{P}_u$ is always diagonal, and impulse response matrix $\mathbf{W}$ can be written in column-partitioned form, as

$$\mathbf{W} = (\mathbf{w}_1 \mid \mathbf{w}_2 \mid \dots \mid \mathbf{w}_N) . \tag{7-50}$$

It is more convenient to begin with the transpose of $\mathbf{u}^{MV}(N)$ rather than $\mathbf{u}^{MV}(N)$. Observe, from Eq. (7-49), that

$$[\mathbf{u}^{MV}(N)]' = \mathbf{z}'(\mathbf{W}\mathbf{P}_u\mathbf{W}' + v_n\mathbf{I})^{-1}\mathbf{W}\mathbf{P}_u ,$$

or,

$$(u^{MV}(1 \mid N), u^{MV}(2 \mid N), \dots , u^{MV}(k \mid N), \dots , u^{MV}(N \mid N))'$$
$$= \mathbf{z}'(\mathbf{W}\mathbf{P}_u\mathbf{W}' + v_n\mathbf{I})^{-1}(\mathbf{w}_1 \mid \mathbf{w}_2 \mid \dots \mid \mathbf{w}_k \mid \dots \mid \mathbf{w}_N)\mathbf{P}_u ;$$

so that

$$(u^{MV}(1 \mid N)v_u^{-1}(1), u^{MV}(2 \mid N)v_u^{-1}(2), \dots ,$$
$$u^{MV}(k \mid N)v_u^{-1}(k), \dots , u^{MV}(N \mid N)v_u^{-1}(N))'$$
$$= \mathbf{z}'(\mathbf{W}\mathbf{P}_u\mathbf{W}' + v_n\mathbf{I})^{-1}(\mathbf{w}_1 \mid \mathbf{w}_2 \mid \dots \mid \mathbf{w}_k \mid \dots \mid \mathbf{w}_N) ; \tag{7-51}$$

hence,

$$u^{MV}(k \mid N) = \mathbf{z}'(\mathbf{W}\mathbf{P}_u\mathbf{W}' + v_n\mathbf{I})^{-1}\mathbf{w}_k v_u(k) ,$$

or, because $u^{MV}(k \mid N)$ is a scalar (so that its transpose equals itself),

$$u^{MV}(k \mid N) = v_u(k)\mathbf{w}'_k(\mathbf{W}\mathbf{P}_u\mathbf{W}' + v_n\mathbf{I})^{-1}\mathbf{z} . \tag{7-52}$$

Another result that we will need is the covariance matrix of $\mathbf{u}_{ERR}(N)$ and its k th component. From the basic definition of such a matrix, we find that

$$\text{Covar}[\mathbf{u}_{ERR}(N)] = \mathbf{E}\{[\mathbf{u}_{ERR}(N) - \mathbf{E}\{\mathbf{u}_{ERR}(N)\}][\mathbf{u}_{ERR}(N) - \mathbf{E}\{\mathbf{u}_{ERR}(N)\}]'\}$$
$$= \mathbf{E}\{\mathbf{u}_{ERR}(N)\mathbf{u}_{ERR}(N)'\} ,$$

because, as we have shown in Eq. (7-41), $\mathbf{u}_{ERR}(N)$ is zero mean; hence,

$$\text{Covar}[\mathbf{u}_{ERR}(N)] = \mathbf{E}\{[\mathbf{u} - \mathbf{u}^{MV}(N)][\mathbf{u} - \mathbf{u}^{MV}(N)]'\}$$
$$= \mathbf{E}\{[\mathbf{u} - \mathbf{P}_u\mathbf{W}'(\mathbf{W}\mathbf{P}_u\mathbf{W}' + v_n\mathbf{I})^{-1}\mathbf{z}]$$
$$[\mathbf{u} - \mathbf{P}_u\mathbf{W}'(\mathbf{W}\mathbf{P}_u\mathbf{W}' + v_n\mathbf{I})^{-1}\mathbf{z}]'\}$$
$$= \mathbf{P}_u - \mathbf{P}_u\mathbf{W}'(\mathbf{W}\mathbf{P}_u\mathbf{W}' + v_n\mathbf{I})^{-1}\mathbf{P}_{zu}$$
$$- \mathbf{P}_{uz}(\mathbf{W}\mathbf{P}_u\mathbf{W}' + v_n\mathbf{I})^{-1}\mathbf{W}\mathbf{P}_u$$
$$+ \mathbf{P}_u\mathbf{W}'(\mathbf{W}\mathbf{P}_u\mathbf{W}' + v_n\mathbf{I})^{-1}\mathbf{P}_z(\mathbf{W}\mathbf{P}_u\mathbf{W}' + v_n\mathbf{I})^{-1}\mathbf{W}\mathbf{P}_u .$$

Substituting Eqs. (7-47) and (7-48) for $\mathbf{P}_{uz}$ and $\mathbf{P}_z$, respectively, and using the fact that $\mathbf{P}_{zu} = [\mathbf{P}_{uz}]'$, we find that

$$\text{Covar}[\mathbf{u}_{ERR}(N)] = \mathbf{P}_u - \mathbf{P}_u\mathbf{W}'\,(\mathbf{W}\mathbf{P}_u\mathbf{W}' + v_n\mathbf{I})^{-1}\mathbf{W}\mathbf{P}_u\,, \qquad (7\text{-}53)$$

which is the desired result.

Working with the partitioned version of matrix $\mathbf{W}$, it is easy to show that

$$\text{Var}[\mathbf{u}_{ERR}(k \mid N)] = v_u(k) - v_u^2(k)\mathbf{w}'_k(\mathbf{W}\mathbf{P}_u\mathbf{W}' + v_n\mathbf{I})^{-1}\mathbf{w}_k\,. \qquad (7\text{-}54)$$

Two special cases merit examination.

Case 1. Event locations (i.e., Q) are unknown. In this case,

$$\mathbf{P}_u = \mathbf{E}\{\mathbf{u}\mathbf{u}'\} = \mathbf{E}\{(\mathbf{Q}\mathbf{r} + \mathbf{u}_B)(\mathbf{Q}\mathbf{r} + \mathbf{u}_B)'\} = (\lambda v_r + v_B)\mathbf{I}\,. \qquad (7\text{-}55)$$

In deriving this result we have used the fact [recall, from probability theory (e.g., Papoulis, 1984), that $\mathbf{E}\{ab\} = \mathbf{E}\{a\,\mathbf{E}\{b \mid a\}\}$, where the outer expectation is with respect to a] that

$$\mathbf{E}\{\mathbf{Q}\mathbf{r}\mathbf{r}'\mathbf{Q}\} = \mathbf{E}_Q\{\mathbf{Q}\mathbf{E}\{\mathbf{r}\mathbf{r}' \mid \mathbf{Q}\}\mathbf{Q}\} = v_r\mathbf{E}_Q\{\mathbf{Q}^2\} = v_r\mathbf{E}_Q\{\mathbf{Q}\} = v_r\lambda\,.$$

Note that this is the case that is associated with detection, and for it, $v_u(k) = v_u = \lambda v_r + v_B$.

Case 2. Event locations (i.e., Q) are known. In this case,

$$\mathbf{P}_u = \mathbf{E}\{\mathbf{u}\mathbf{u}'\} = \mathbf{E}\{(\mathbf{Q}\mathbf{r} + \mathbf{u}_B)(\mathbf{Q}\mathbf{r} + \mathbf{u}_B)'\} = v_r\mathbf{Q} + v_B\mathbf{I}\,. \qquad (7\text{-}56)$$

In deriving this result we have used the fact, that

$$\mathbf{E}\{\mathbf{Q}\mathbf{r}\mathbf{r}'\mathbf{Q}\} = \mathbf{Q}\mathbf{E}\{\mathbf{r}\mathbf{r}'\}\mathbf{Q} = v_r\mathbf{Q}^2 = v_r\mathbf{Q}\,.$$

Substituting Eq. (7-56) into Eq. (7-49), we find that

$$\mathbf{u}^{MV}(N \mid \mathbf{q}) = (v_r\mathbf{Q} + v_B\mathbf{I})\mathbf{W}'(v_r\mathbf{W}\mathbf{Q}\mathbf{W}' + v_B\mathbf{W}\mathbf{W}' + v_n\mathbf{I})^{-1}\mathbf{z}\,. \qquad (7\text{-}57)$$

This looks very familiar. When $\mathbf{a}, \mathbf{b}, \mathbf{s}$, and $\mathbf{q}$ are all known, then the *maximum-likelihood values* of $\mathbf{r}$ and $\mathbf{u}_B$, which are given in Eqs. (7-29) and (7-30), respectively, are

$$\mathbf{r}^{ML} = \mathbf{r}^{ML}(N \mid \mathbf{q}) = v_r\mathbf{Q}\mathbf{W}'(v_r\mathbf{W}\mathbf{Q}\mathbf{W}' + v_B\mathbf{W}\mathbf{W}' + v_n\mathbf{I})^{-1}\mathbf{z} \qquad (7\text{-}58)$$

and

$$\mathbf{u}_B{}^{ML} = \mathbf{u}_B{}^{ML}(N \mid \mathbf{q}) = v_B\mathbf{W}'(v_r\mathbf{W}\mathbf{Q}\mathbf{W}' + v_B\mathbf{W}\mathbf{W}' + v_n\mathbf{I})^{-1}\mathbf{z}\,. \qquad (7\text{-}59)$$

We have arrived at the very important connection between MLD and MVD, that *when* **a**, **b**, **s**, *and* **q** *are all known*, then

$$\mathbf{u}^{MV}(N \mid \mathbf{q}) = \mathbf{r}^{ML}(N \mid \mathbf{q}) + \mathbf{u}_B{}^{ML}(N \mid \mathbf{q}) \tag{7-60}$$

When **a**, **b**, **s**, and **q** are known then all sources of uncertainty in the convolutional model are Gaussian, and in this case it is a well known fact (e.g., Sorenson, 1980, Mendel, 1987), in the theory of random parameter estimation, that the unconditional maximum-likelihood and minimum-variance estimators of a random parameter vector are the same. This is obvious in our problem, when we compare Eqs. (7-60) and (7-33). Consequently,

$$\mathbf{u}^{MV}(N \mid \mathbf{q}) = \mathbf{r}^{MV}(N \mid \mathbf{q}) + \mathbf{u}_B{}^{MV}(N \mid \mathbf{q}) \tag{7-61}$$

7.5 Threshold Detector

Here we shall show that *the threshold detector decision strategy is*:

$$\textit{If } [u^{MV}(k \mid N)]^2 > t(k) \textit{ decide } q(k) = 1$$
$$\textit{If } [u^{MV}(k \mid N)]^2 < t(k) \textit{ decide } q(k) = 0 , \tag{7-62}$$

where

$$t(k) = \{A_0(k)A_1(k)/[A_1(k) - A_0(k)]\} \{\ln[A_1(k)/A_0(k)]$$
$$- 2\ln[\lambda/(1 - \lambda)]\} \tag{7-63}$$

and, for q = 0 *and* 1,

$$A_q(k) = \{1 - \text{Var}[u_{ERR}(k \mid N)]/v_u\}^2[qv_r + v_B]$$
$$+ \text{Var}[u_{ERR}(k \mid N)]\{1 - \text{Var}[u_{ERR}(k \mid N)]/v_u\} . \tag{7-64}$$

Proof [Adapted from Mendel (1983) and Kormylo (1979)]: The starting point for the derivation of the Threshold Detector is $u^{MV}(k \mid N)$, where, as we have seen,

$$u^{MV}(k \mid N) = v_u(k)\mathbf{w}_k{}'(\mathbf{WP_uW'} + v_n\mathbf{I})^{-1}\mathbf{z} . \tag{7-65}$$

To begin, we shall express $u^{MV}(k \mid N)$ as a function of u(k). This will be a traditional *signal plus noise model*. For notational convenience, let

$$\mathbf{f}'(k) = v_u(k)\mathbf{w}_k{}'(\mathbf{WP_uW'} + v_n\mathbf{I})^{-1} \tag{7-66}$$

so that $u^{MV}(k \mid N)$ can be expressed as

$$u^{MV}(k \mid N) = f'(k)z .$$ (7-67)

Next substitute the convolutional model,

$$z = Wu + n ,$$

and the partitioned form of W into Eq. (7-67), to see that

$$u^{MV}(k \mid N) = f'(k)(Wu + n) = f'(k)[w_1 \mid w_2 \mid \dots \mid w_N]u + f'(k)n$$
$$u^{MV}(k \mid N) = [f'(k)w_1 , f'(k)w_2 , \dots , f'(k)w_N]col[u(1),u(2), \dots ,u(N)]$$
$$\qquad\qquad + f'(k)n$$
$$u^{MV}(k \mid N) = f'(k)w_k u(k) + e(k) ,$$ (7-68)

where

$$e(k) = [f'(k)w_1, f'(k)w_2, \dots , f'(k)w_{k-1}, 0, f'(k)w_{k+1}, \dots , f'(k)w_N]u$$
$$\qquad + f'(k)n.$$ (7-69)

Again, for notational simplicity, let

$$c(k) = f'(k)w_k ,$$ (7-70)

so that

$$u^{MV}(k \mid N) = c(k)u(k) + e(k).$$ (7-71)

We have indeed been able to express the estimate of $u(k)$ as an explicit function of the desired signal, $u(k)$. In Eq. (7-71) $u^{MV}(k \mid N)$, which is obtained from an MVD filter, is treated as a known measurement.

From this point on, our development of the threshold detector follows the derivation given in Mendel (1983, pp. 127-130) (see, also, Kormylo, 1979).

Observe, from Eq. (7-69), that $e(k)$ is independent of $u(k)$ and is of zero mean. Now we come to the major assumption in the derivation of the threshold detector. *We assume that* $e(k)$ *is Gaussian.* Of course this is an incorrect assumption in the seismic case, where $u(k)$ is the sum of Bernoulli-Gaussian and backscatter processes. That is why other detectors that start where the threshold detector leaves off will do a better job than the threshold detector. We will see that the threshold detector only uses the first-and second-order statistics of $e(k)$; hence, we can think of a Gaussian $e(k)$ as being statistically equivalent, through the first two moments, to the actual $e(k)$.

We use the following *unconditional likelihood function* for detecting $q(k)$, given $u^{MV}(k \mid N)$ (because a, b, and s are assumed given, we suppress the dependence of the likelihood function on them):

$$\mathcal{S}\{q(k) \mid u^{MV}(k \mid N)\} = p[u^{MV}(k \mid N) \mid q(k)]Pr[q(k)] .$$ (7-72)

Note that conditioning with respect to $u^{MV}(k \mid N)$ is the same as conditioning with respect to z, because $u^{MV}(k \mid N)$ is a function of z. Consequently, $\mathcal{S}$ is linked closely to $\mathcal{M}$ or $\mathcal{N}$, but is different from these functions.

The detector that is based on $\mathcal{S}$ is going to provide us with an *initial* q sequence for our Block Component Method; hence, this sequence can be found by solving an optimization problem that does not have to be compatible with the original one that is associated with $\mathcal{M}$ or $\mathcal{N}$. Our recursive detectors, that begin with this initial q sequence,will be compatible with $\mathcal{M}$ or $\mathcal{N}$.

Examining Eq. (7-71), we see that, under the assumption that e(k) is Gaussian, and, when q(k) is fixed [in which case, u(k) is Gaussian], then $u^{MV}(k \mid N)$ is Gaussian, i.e., $p[u^{MV}(k \mid N) \mid q(k)]$ is a Gaussian density function of the form

$$p[u^{MV}(k \mid N) \mid q(k)] = [2\pi A_q(k)]^{-1/2} \exp\{- [u^{MV}(k \mid N)]^2/2A_q(k)\} , \quad (7\text{-}73)$$

in which

$$A_q(k) = E\{[u^{MV}(k \mid N)]^2 \mid q(k) = q]\} \quad (7\text{-}74)$$

and q = 0 or 1. The *decision strategy* that leads to the threshold detector is the following unconditional maximum-likelihood detection rule:

$$\text{If } \ln D(z;k) \geqslant 0 \text{ } \textit{decide } q(k) = 1;$$
$$\text{If } \ln D(z;k) \leqslant 0 \text{ } \textit{decide } q(k) = 0, \quad (7\text{-}75)$$

where D(z;k) is the following *unconditional likelihood ratio*,

$$\begin{aligned} D(z;k) &= \mathcal{S}\{q(k) = 1 \mid u^{MV}(k \mid N)\}/\mathcal{S}\{q(k) = 0 \mid u^{MV}(k \mid N)\} \\ &= p[u^{MV}(k \mid N) \mid q(k) = 1]Pr[q(k) = 1]/ \\ &\quad p[u^{MV}(k \mid N) \mid q(k) = 0] \, Pr[q(k) = 0] . \end{aligned} \quad (7\text{-}76)$$

From this equation, it is easy to see that

$$\begin{aligned} \ln D(z;k) &= \ln\{p[u^{MV}(k \mid N) \mid q(k) = 1]/p[u^{MV}(k \mid N) \mid q(k) = 0]\} \\ &\quad + \ln\{Pr[q(k) = 1]/Pr[q(k) = 0]\} . \end{aligned} \quad (7\text{-}77)$$

Now

$$\ln\{Pr[q(k) = 1]/Pr[q(k) = 0]\} = \ln[\lambda/(1 - \lambda)] , \quad (7\text{-}78)$$

and [from Eq. (7-73)]

$$\begin{aligned} \ln\{p[u^{MV}(k \mid N) \mid q(k) &= 1]/p[u^{MV}(k \mid N) \mid q(k) = 0]\} \\ &= \tfrac{1}{2}\ln[A_0(k)/A_1(k)] \\ &\quad + \tfrac{1}{2} [u^{MV}(k \mid N)]^2\{[A_0(k)]^{-1} - [A_1(k)]^{-1}\} , \quad (7\text{-}79) \end{aligned}$$

where A_0 and A_1 are short for $A_{q=0}$ and $A_{q=1}$, respectively. Substituting Eqs. (7-78) and (7-79) into Eq. (7-77), we find that

$$\text{Ln}D(z;k) = \frac{1}{2} [u^{MV}(k \mid N)]^2 \{[A_0(k)]^{-1} - [A_1(k)]^{-1}\}$$
$$+ \frac{1}{2} \text{Ln}[A_0(k)/A_1(k)] + \text{Ln}[\lambda/(1 - \lambda)] . \tag{7-80}$$

Our unconditional maximum-likelihood detection rule, given in Eq. (7-75), can now be expressed as the following threshold test on $[u^{MV}(k \mid N)]^2$:

If $[u^{MV}(k \mid N)]^2 \geqslant \{A_0(k)A_1(k)/[A_1(k) - A_0(k)]\}\{\text{Ln}[A_1(k)/A_0(k)]$
$\qquad - 2\text{Ln}[\lambda/(1 - \lambda)]\}$ *decide* q(k) = 1
If $[u^{MV}(k \mid N)]^2 \leqslant \{A_0(k)A_1(k)/[A_1(k) - A_0(k)]\}\{\text{Ln}[A_1(k)/A_0(k)]$
$\qquad - 2\text{Ln}[\lambda/(1 - \lambda)]\}$ *decide* q(k) = 0. $\tag{7-81}$

Of course, in order to compute the threshold function on the right-hand side of this equation, we must be able to compute $A_0(k)$ and $\ddot{A}_1(k)$. We must now evaluate $A_q(k)$ in Eq. (7-74), remembering, from Chapter 2, that

$$u(k) = r(k)q(k) + u_B(k). \tag{7-82}$$

Substituting Eqs. (7-71) and (7-82) into Eq. (7-74), we find that

$$A_q(k) = E\{[c^2(k)u^2(k) + 2c(k)u(k)e(k) + e^2(k)] \mid q(k) = q\}$$
$$= c^2(k)E\{[r(k)q(k) + u_B(k)]^2 \mid q(k) = q\} + E\{e^2(k) \mid q(k) = q\} ,$$

because e(k) and u(k) are statistically independent. Consequently,

$$A_q(k) = c^2(k)E\{[r(k)q + u_B(k)]^2\} + E\{e^2(k)\} \tag{7-83}$$

because e(k) does not depend on q(k). Continuing, we find that

$$A_q(k) = c^2(k)[qv_r + v_B] + E\{e^2(k)\} \tag{7-84}$$

because $q^2 = q$ and r(k) and $u_B(k)$ are statistically independent.

Next, we shall compute $E\{e^2(k)\}$. To do this we shall compute another formula for $\text{Var}[u_{ERR}(k \mid N)]$, a quantity that is available from the MVD filter. Recall that

$$u_{ERR}(k \mid N) = u(k) - u^{MV}(k \mid N) .$$

Substituting Eq.(7-71) into this expression, we find that

$$u_{ERR}(k \mid N) = [1 - c(k)]u(k) - e(k);$$

hence,

$$\text{Var}[u_{ERR}(k \mid N)] = [1 - c(k)]^2 v_u + E\{e^2(k)\} ,$$

or,

$$E\{e^2(k)\} = \text{Var}[u_{ERR}(k \mid N)] - [1 - c(k)]2v_u , \qquad (7\text{-}85)$$

which is a computible formula for $E\{e^2(k)\}$, i.e., everything on the right-hand side of Eq. (7-85) can be computed. Note that

$$v_u = E\{u^2(k)\} = v_r\lambda + v_B . \qquad (7\text{-}86)$$

Substituting Eq. (7-85) into Eq. (7-84), we obtain the following expression for $A_q(k)$

$$A_q(k) = c^2(k)[qv_r + v_B] + \text{Var}[u_{ERR}(k \mid N)] - [1 - c(k)]^2 v_u . \qquad (7\text{-}87)$$

This equation is in terms of both $c(k)$ and $\text{Var}[u_{ERR}(k \mid N)]$.

Next, we show how to express $A_q(k)$ just as a function of $\text{Var}[u_{ERR}(k \mid N)]$. From Eqs. (7-70), (7-66) and (7-54), it is easy to show that

$$c(k) = 1 - \text{Var}[u_{ERR}(k \mid N)]/v_u . \qquad (7\text{-}88)$$

Substituting Eq. (7-88) into Eq. (7-87), we obtain the final result, that

$$\begin{aligned}A_q(k) = &\{1 - \text{Var}[u_{ERR}(k \mid N)]/v_u\}^2 [qv_r + v_B] \\ &+ \text{Var}[u_{ERR}(k \mid N)]\{1 - \text{Var}[u_{ERR}(k \mid N)]/v_u\} .\end{aligned} \qquad (7\text{-}89)$$

This completes the derivation of the Threshold Detector.

Observe how the threshold detector uses both the MVD output, $u^{MV}(k \mid N)$, and its error variance, $\text{Var}[u_{ERR}(k \mid N)]$. Because of the squaring up of $u^{MV}(k \mid N)$, the threshold detector is also known as a *square-law detector*. Once again, the reader is reminded that we have assumed that a, b, and s are known. These known values are used to obtain both $u^{MV}(k \mid N)$ and $\text{Var}[u_{ERR}(k \mid N)]$ from the MVD filter. When we run that filter we must set v_u equal to its value computed in Eq. (7-86). The threshold detector is summarized in Figure 7-1.

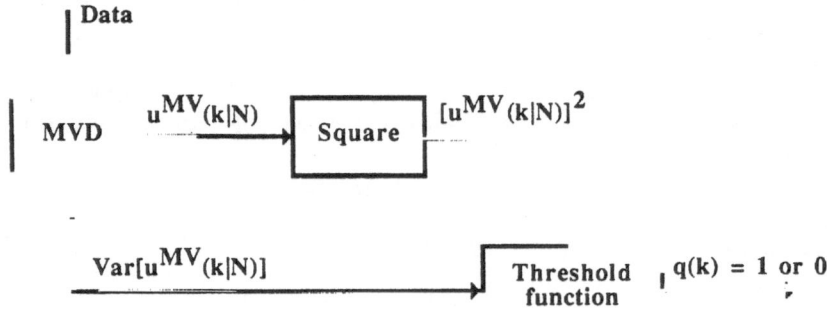

Figure 7-1. Threshold detector and its operations.

7.6 Single Most-Likely Replacement Detector

A complete derivation of the SMLR detector can be found in Mendel (1983, pp. 131-134) [or in Kormylo ,1979, or Kormylo and Mendel, 1982]. Here we shall only provide a brief outline of that derivation, and shall summarize the SMLR decision function.

We use the functions $\mathcal{M}\{q_{t,k} \mid z\}$ and $\mathcal{M}\{q_r \mid z\}$ for choosing between the test and reference q, $q_{t,k}$ and q_r, respectively. These functions are defined in Equation (7-21), or Equation (7-3). We have suppressed the dependence of $\mathcal{M}$ on a, b, and s, because in detection these parameters are all assumed to be known. The associated exponentials of these functions are denoted $M\{q_{t,k} \mid z\}$ and $M\{q_r \mid z\}$, where

$$M = \exp \mathcal{M} . \qquad (7\text{-}90)$$

Let $D(z;k)$ denote the SMLR decision function. It is defined as

$$D(z;k) = M\{q_{t,k} \mid z\}/M\{q_r \mid z\} ; \qquad (7\text{-}91)$$

hence,

$$\ln D(z;k) = \ln M\{q_{t,k} \mid z\} - \ln M\{q_r \mid z\} = \mathcal{M}\{q_{t,k} \mid z\} - \mathcal{M}\{q_r \mid z\}. \qquad (7\text{-}92)$$

Our detection rule for choosing between sequences $q_{t,k}$ and q_r is given by

$$\text{If } \ln D(z;k) > 0 \text{ choose } q_{t,k} \text{ over } q_r ;$$
$$\text{If } \ln D(z;k) < 0 \text{ choose } q_r \text{ over } q_{t,k} . \qquad (7\text{-}93)$$

If $\ln D(z;k) = 0$, then q_r is chosen. The *SMLR detector decision strategy* is to examine all the values of k for which $\ln D(z;k) > 0$ and to find the value of k at which $\ln D(z;k)$ is a maximum. This time point, which we designate k', is then the single time point at which a change is made in our reference sequence.

Recall that $q_{t,k}$ and q_r differ at only one time point, the k th, i.e.,

$$q_{t,k}(i) = q_r(i) \text{ for all } i \neq k$$
$$q_{t,k}(i) = 1 - q_r(i) \text{ for } i = k. \qquad (7\text{-}94)$$

Substituting Equation (7-3) into (7-92), it is possible to show (with a significant amount of effort) that

$$\ln D(z;k) = (w_k'P_z^{-1}z)^2/\{v_r^{-1}[q_{t,k}(k) - q_r(k)]^{-1} + w_k'P_z^{-1}w_k\}$$
$$+ 2[q_{t,k}(k) - q_r(k)]\ln[\lambda/(1 - \lambda)] . \qquad (7\text{-}95)$$

Note that somewhat different results are obtained when loglikelihood function $\mathcal{N}$ is used instead of function $\mathcal{M}$. See Chapter 8 for a brief discussion about this.

The SMLR detector search algorithm is summarized in Figure 7-2. The "information" used by the SMLR detector is contained in $\mathbf{a}$, $\mathbf{b}$, and $\mathbf{s}$, all of which are assumed known. The "linear processor" generates the two scalars, $\mathbf{w}_k'\mathbf{P}_z^{-1}\mathbf{z}$ and $\mathbf{w}_k'\mathbf{P}_z^{-1}\mathbf{w}_k$, which are needed to compute $\ln D(\mathbf{z};k)$ using (7-95). We demonstrate next that these quantities are obtained from MVD filtering.

Referring to Eqs. (7-36) and (7-38), and using Eq. (7-48), we see that $u^{MV}(k\,|\,N)$ and $Var[u^{ERR}(k\,|\,N)]$ can be expressed as

$$u^{MV}(k\,|\,N) = v_u(k)\mathbf{w}_k'\mathbf{P}_z^{-1}\mathbf{z} \qquad (7\text{-}96)$$

and

$$Var[u_{ERR}(k\,|\,N)] = v_u(k) - v_u^2(k)\mathbf{w}_k'\mathbf{P}_z^{-1}\mathbf{w}_k \; . \qquad (7\text{-}97)$$

From these equations, it is clear that

$$\mathbf{w}_k'\mathbf{P}_z^{-1}\mathbf{z} = u^{MV}(k\,|\,N)/v_u(k) \qquad (7\text{-}98)$$

and

$$\mathbf{w}_k'\mathbf{P}_z^{-1}\mathbf{w}_k = \{v_u(k) - Var[u_{ERR}(k\,|\,N)]\}/v_u^2(k) \; . \qquad (7\text{-}99)$$

MVD is, therefore, an essential ingredient to SMLR detection.

As noted in Chapter 4, the SMLR detector is not self-starting. Initial reference sequence $\mathbf{q}_r$ must be obtained from some other detection method, or, perhaps it is provided by some interpretive technique, such as looking at the data and guessing bands of time points at which it appears to be highly likely that events are present. A threshold detector can be used in the former situation.

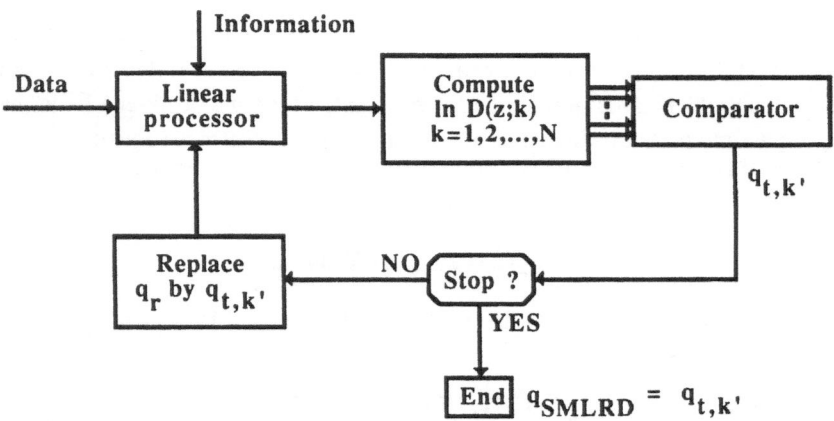

Figure 7-2. Single most likely replacement detector search algorithm.

7.7 Single Spike Shift Detector

Chi and Mendel (1984a) state that "The SMLR detector is derived by assuming that q_r and q_t differ in just one location. In order to detect a better q sequence, i.e., one with a higher likelihood function than that obtained by the SMLR detector, we shall derive a detector that computes the likelihood ratio of q_t and q_r when q_t and q_r differ at exactly two locations.

"The total number of q sequences obtained from q_r by changing any two locations is $N(N-1)/2$, which can be a very large number. We shall only consider the special case where q_t and q_r differ at two *consecutive* locations. The *SSS detector restricts* q_t *to be generated from* q_r *by shifting only at those k where there is a spike in* q_r. *Such spikes are shifted one location forwards or backwards.*"

Reference sequence q_r is always composed of two uniquely different collections of unit spikes, namely *isolated spikes* and *clumps of spikes*. This is illustrated in Figure 7-3. Observe that there are 5 isolated spikes, located at $k = 15, 18, 30, 40$ and 45, and three clumps of spikes. Each clump can contain a different number of spikes, but it must contain at least two adjacent spikes to be called a clump. The first clump contains 4 spikes located at $k = 4, 5, 6$, and 7; the second clump contains 3 spikes located at $k = 21, 22$, and 23; and, the third clump contains just two spikes, located at $k = 34$ and 35. Note that a spike at k_i is said to "isolated" if $q(k_i) = 1$ and $q(k_i + 1) = q(k_i - 1) = 0$.

We shall assume that q_r includes L isolated spikes and G clumps of spikes. The L isolated spikes are located at $k = k_1, k_2, \ldots, k_L$. Each clump can have a different number of spikes in it. The i th clump contains g_i spikes; its first spike occurs at $k = l_i$; and, its last spike occurs at $k = l_i + g_i - 1$. Note that $i = 1, 2, \ldots, G$.

In Chapter 4 we noted that (see Figure 4-14) shifting a unit spike can cause a change in the reference sequence to occur at either one or two locations. The L isolated spikes and G clumps of spikes can be decomposed into two mutually exclusive subclasses C_1 and C_2. C_1 contains those spikes which, when shifted either one unit to the right or left from their present positions, cause a change in the reference sequence to occur at exactly one location. C_2 contains those spikes which, when similarly shifted, cause a change in the reference sequence to occur

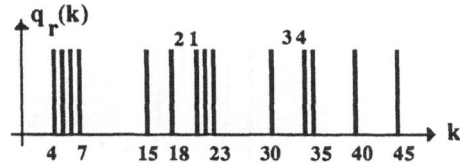

Figure 7-3. A reference sequence that contains isolated spikes and clumps of spikes.

at exactly two locations. Which elements are in C_1 and C_2?

All isolated spikes are in C_2. This is demonstrated in Figure 7-4. When the spike at $k = k_i$ is shifted one unit to the left a change occurs at the two time points $k = k_i - 1$ and $k = k_i$. When the spike at $k = k_i$ is shifted one unit to the right a change occurs at the two time points $k = k_i$ and $k = k_i + 1$.

All interior points of a clump are in C_1. This is demonstrated in Figure 7-5. The clump of 5 spikes has 3 interior points, located at $k = l_i + 1$, $l_i + 2$, and $l_i + 3$. When, for example, the interior spike at $k = l_i + 2$ is shifted either one unit to the left or right, only one change occurs, at the location of that spike.

End points of a clump are in C_1 *when they are shifted into the clump, and they are in* C_2 *when they are shifted away from the clump.* This is demonstrated in Figure 7-6.

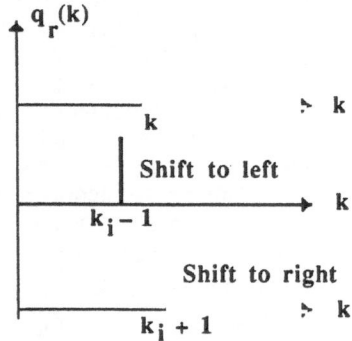

Figure 7-4. An isolated spike, and, what happens to the reference sequence when that spike is shifted either one unit to the left or right.

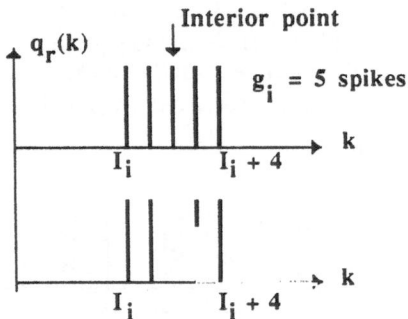

Figure 7-5. A clump of spikes, and what happens to the reference sequence when its interior points are shifted either one unit to the right or left.

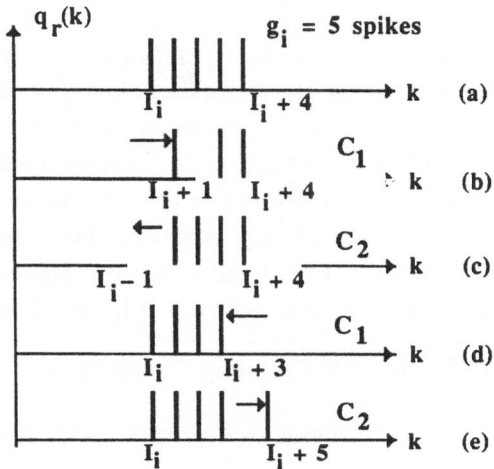

Figure 7-6. End points of a clump of spikes are in C_1 when they are shifted into the clump, as in (b) and (d), and they are in C_2 when they are shifted away from the clump, as in (c) and (e).

To summarize:

$$C_1 = \{q_t \mid q_t(j) = q_r(j) \text{ for all } j \neq k, \text{ and } q_t(j) = 1 - q_r(j) \text{ for all } j = k,$$
where k is an element of the set S_1} (7-100)

and

$$S_1 = \{k \mid l_i + 1 \leqslant k \leqslant l_i + g_i - 2, \text{ for all } 1 \leqslant i \leqslant G, \text{ and, } k = l_i, l_i + g_i - 1$$
when shifted into the clump, for all $1 \leqslant i \leqslant G\}$; (7-101)

additionally,

$$C_2 = \{q_t \mid q_t(j) = q_r(j) \text{ for all } j \neq k, k + 1, \text{ and } q_t(j) = 1 - q_r(j)$$
for all $j = k, k + 1$, where k is an element of the set S_2} (7-102)

and

$$S_2 = \{k \mid k = k_i, k_i - 1 \text{ for all } 1 \leqslant i \leqslant L, \text{ and, } k = l_i - 1 \text{ and}$$
$k = l_i + g_i - 1$ for all $1 \leqslant i \leqslant G\}$. (7-103)

Set S_2 may appear a bit mysterious, but it is really not. When $k = k_i$, then its C_2 sequence differs from the reference sequence at the two points $j = k_i$

and $j = k_i + 1$. This corresponds to the case when the isolated spike at $k = k_i$ is moved one unit to the right. When $k = k_i - 1$, then its C_2 sequence differs from the reference sequence at the two points $j = k_i - 1$ and $j = k_i$. This corresponds to the case when the isolated spike at $k = k_i$ is moved one unit to the left. When $k = l_i - 1$, then its C_2 sequence differs from the reference sequence at the two points $j = l_i - 1$ and $j = l_i$. This corresponds to the case when the clump's left endpoint spike at $k = l_i$ is moved one unit to the left. Finally, when $k = l_i + g_i - 1$, then its C_2 sequence differs from the reference sequence at the two points $j = l_i + g_i - 1$ and $j = l_i + g_i$. This corresponds to the case when the clump's right endpoint spike at $k = l_i + g_i - 1$ is moved one unit to the right.

The total number of elements in C_1 and C_2, E_1 and E_2, are

$$E_1 = g_1 + g_2 + \dots + g_G \tag{7-104}$$

and

$$E_2 = 2L + 2G. \tag{7-105}$$

Note that the result in Equation (7-104) follows because the two components that make up S_1 contribute $g_1 + g_2 + \dots + g_G - 2G$ and $2G$ elements, respectively, to E_1. In Chapter 4 we stated that 2M test sequences are created during each iteration of the SSS detector. Clearly,

$$2M = E_1 + E_2. \tag{7-106}$$

As in the SMLR detector, the SSS detector uses the functions $\mathcal{M}\{q_{t,k} \mid z\}$ and $\mathcal{M}\{q_r \mid z\}$ for choosing between $q_{t,k}$ and q_r. Recall that in the SMLR detector $q_{t,k}$ differs from q_r at exactly one point. We have just seen that all of the SSS test sequences that are in subclass C_1 share this same property; hence, *for all points in C_1 we compute the SSS decision function using Eq.(7-95).*

On the other hand, for all points in C_2 we must derive a different SSS decision function. The details of this derivation are found in Chi and Mendel (1984, pg.434-435; these results are for $\mathcal{N}$, whereas ours are for $\mathcal{M}$). Here we state the final result.

Assume that $q_{t,k}$ and q_r differ at two consecutive locations, i.e.,

$$q_{t,k}(i) = q_r(i) \text{ for all } i \neq k \text{ and } k + 1$$
$$q_{t,k}(i) = 1 - q_r(i) \text{ for all } i = k \text{ and } k + 1. \tag{7-107}$$

For this case

$$
\begin{aligned}
2\ln D_1(z;k) = v_r \{ & [q_{t,k}(k) - q_r(k)](w_k'\mathbf{P_z}^{-1}z)^2 \{1 + v_r[q_{t,k+1}(k+1) \\
& - q_r(k+1)]w_{k+1}'\mathbf{P_z}^{-1}w_{k+1}\} + [q_{t,k+1}(k+1) \\
& - q_r(k+1)](w_{k+1}'\mathbf{P_z}^{-1}z)^2 \{1 + v_r[q_{t,k}(k) \\
& - q_r(k)]w_k'\mathbf{P_z}^{-1}w_k\} - 2v_r[q_{t,k}(k) - q_r(k)][q_{t,k+1}(k+1) \\
& - q_r(k+1)](w_k'\mathbf{P_z}^{-1}w_{k+1})(w_k'\mathbf{P_z}^{-1}z)(w_{k+1}'\mathbf{P_z}^{-1}z)\}/|A| \\
& + 2[q_{t,k}(k) - q_r(k) + q_{t,k+1}(k+1) \\
& - q_r(k+1)]\ln[\lambda/(1-\lambda)]
\end{aligned}
\tag{7-108}
$$

where

$$
A = \begin{pmatrix} A_{11} & A_{12} \\ \\ A_{21} & A_{22} \end{pmatrix}
$$

and

$$
A_{11} = 1 + v_r[q_{t,k}(k) - q_r(k)]w_k'\mathbf{P_z}^{-1}w_k ,
\tag{7-109}
$$

$$
A_{12} = v_r[q_{t,k+1}(k+1) - q_r(k+1)]w_k'\mathbf{P_z}^{-1}w_{k+1} ,
\tag{7-110}
$$

$$
A_{21} = v_r[q_{t,k}(k) - q_r(k)]w_k'\mathbf{P_z}^{-1}w_{k+1} ,
\tag{7-111}
$$

and,

$$
A_{22} = 1 + v_r[q_{t,k+1}(k+1) - q_r(k+1)]w_{k+1}'\mathbf{P_z}^{-1}w_{k+1} .
\tag{7-112}
$$

Although the decision function in Eq. (7-108) looks awesome, most of the calculations are already available from an MVD filter. This is true of $w_k'\mathbf{P_z}^{-1}w_k$ and $w_k'\mathbf{P_z}^{-1}z$, which can be computed using Eq. (7-99) and (7-98), respectively. It is very easy to obtain similar expressions for $w_{k+1}'\mathbf{P_z}^{-1}w_{k+1}$ and $w_{k+1}'\mathbf{P_z}^{-1}z$. Just increment k to k+1 in the former expressions. The only term that is new is $w_k'\mathbf{P_z}^{-1}w_{k+1}$. A practical method for computing this term is found in Chi and Mendel (1985). Note, however, that

$$
E\{u^{MV}(k \mid N)u^{MV}(k+1 \mid N)\} = v_u(k)v_u(k+1)w_k'\mathbf{P_z}^{-1}w_{k+1} .
\tag{7-113}
$$

This can be obtained by substituting Eq. (7-96), for both $u^{MV}(k \mid N)$ and $u^{MV}(k+1 \mid N)$, into the expectation. From this result we see that the new term $w_k'\mathbf{P_z}^{-1}w_{k+1}$ is related to the crosscorrelation between successive minimum-variance estimates. If we calculate the entire covariance matrix for $u_{ERR}(N)$, using Equation (7-37), then $E\{u^{MV}(k \mid N)u^{MV}(k+1 \mid N)\}$ can be found as the first super-diagonal of that matrix. Of course, it is impractical to carry out this calculation because of the N×N matrix inverse that is needed to compute (7-37).

The *SSS detector decision strategy* is to examine all the values of k for which $\ln D(z;k) > 0$ and $\ln D_1(z;k) > 0$ and to find the value of k at which the

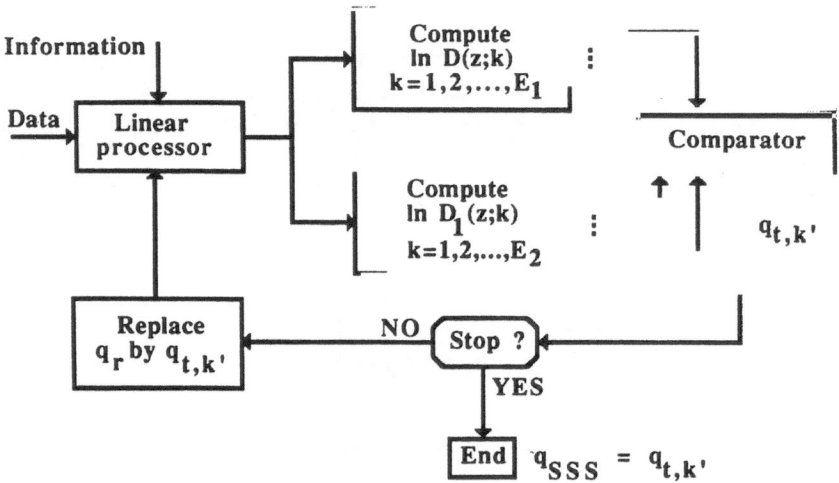

Figure 7-7.   Single spike shift detector search algorithm.

composite function $\{\ln D(z;k) > 0$ and $\ln D_1(z;k) > 0\}$ is a maximum. This time point, which we designate k', is then the single time point at which a change is made in our reference sequence.

The SSS detector search algorithm is summarized in Figure 7-7.

7.8 SSS-SMLR Detector

In Chapter 4 we mentioned that it is possible to combine the SSS and SMLR decision strategies into an SSS-SMLR strategy. Figure 7-7 shows us how to do this. All that needs to be done is to change the number of time points at which the SMLR loglikelihood ratio decision function, $\ln D(z;k)$, is evaluated, from E_1 to N.

The *SSS-SMLR detector decision strategy* is the same as the SSS detector decision strategy. We examine all the values of k for which $\ln D(z;k) > 0$ and $\ln D_1(z;k) > 0$ and find the value of k at which the composite function $\{\ln D(z;k) > 0$ and $\ln D_1(z;k) > 0\}$ is a maximum. This time point, which we designate k', is then the single time point at which a change is made in our reference sequence.

7.9 Marquardt-Levenberg Algorithm

The Marquardt-Levenberg algorithm is a modified Newton Raphson algorithm (Marquardt, 1963). To develop the latter we direct our attention at the function $f(x)$, where f is an arbitrary nonlinear function of a collection of variables that have been assembled into the vector x. The function $f(x)$ is assumed to be continuous and continuously differentiable through its second derivatives with respect to the elements of x.

Let x_i denote the i th approximation to the *maximum* point. To develop the Newton Raphson algorithm we begin by performing a Taylor's series expansion of $f(x)$ about the point x_i. The expansion up to the quadratic terms is

$$f(x) \simeq f(x_i) + [g_x(x_i)]'(x - x_i) + (x - x_i)'H_x(x_i)(x - x_i)/2 , \qquad (7\text{-}114)$$

where $g_x(x_i)$ is the vector of first partial derivatives of f evaluated at x_i, and $H_x(x_i)$ is the matrix of second partial derivatives of f evaluated at x_i. Let the approximation in Eq. (7-114) be designated $f^a(x)$. We may use the maximum of $f^a(x)$ as an approximation to the maximum of $f(x)$. This will be where $\partial f^a(x)/\partial x = 0$. Taking the gradient of $f^a(x)$ with respect to the vector x, and setting the result equal to zero, we find that

$$g_x(x_i) + H_x(x_i)(x - x_i) = 0 , \qquad (7\text{-}115)$$

which can be solved for x, as

$$x = x_i - [H_x(x_i)]^{-1}g_x(x_i) .$$

Designating the left-hand side as the new iterate x_{i+1}, we obtain the Newton Raphson algorithm,

$$x_{i+1} = x_i - [H_x(x_i)]^{-1}g_x(x_i) . \qquad (7\text{-}116)$$

Fox (1971) states "This process can converge extremely rapidly in many problems and is one of the preferred methods of solving systems of simultaneous nonlinear equations. Indeed, if f is a quadratic, we obtain the solution in a single step, since the Taylor series expansion is exact. ... On the other hand, the method may diverge in some problems; it will converge to saddle points and minima ... and sometimes the matrix $H_x(x_i)$ can be singular or near singular in otherwise properly formed problems."

If the matrix $H_x(x_i)$ is not positive definite, so that its inverse is not guaranteed to exist, then it may be made positive definite by adding a large enough diagonal matrix to $H_x(x_i)$. In other words, the algorithm in Eq. (7-116) is modified to

$$x_{i+1} = x_i - [H_x(x_i) + D_i]^{-1}g_x(x_i), \qquad (7\text{-}117)$$

in which D_i is the diagonal matrix. Equation (7-117) is the Marquardt-Levenberg algorithm that is described in Chapter 4.

Of course, to implement this algorithm we must be able to compute the gradient vector $g_x(x_i)$ and the Hessian matrix $H_x(x_i)$. These calculations are problem dependent; so, in the next two sections we describe them for our deconvolution problem.

7.10 Calculating Gradients

In Chapter 4 we saw that there are two functions that can be maximized. They are the original loglikelihood function, $\mathcal{L}\{a, b, s, q, r, u_B \mid z\}$, and the related function, $\mathcal{M}\{a, b, s, q \mid z\}$. Recall that $\mathcal{M}$ was derived from $\mathcal{L}$ in the Separation Theorem. In this section we explain how to compute the gradients of both $\mathcal{M}$ and $\mathcal{L}$ with respect to the wavelet parameters, which are in the vectors a and b, and the variance parameters, which are in s. Because $\mathcal{M}$ and $\mathcal{L}$ depend on these parameters in very different ways, we must treat them separately.

For the convenience of the reader, we restate the formulas for $\mathcal{M}$ and $\mathcal{L}$:

$$\mathcal{M}\{a, b, s, q \mid z\} = -\tfrac{N}{2}\ln(v_n v_r v_B) - \tfrac{1}{2}z'\Omega^{-1}z$$
$$+ m(q)\ln(\lambda) + [N - m(q)]\ln(1 - \lambda) \qquad (7\text{-}118)$$

and

$$\mathcal{L}\{a, b, s, q, r, u_B \mid z\} = -\tfrac{N}{2}\ln(v_n v_r v_B) - r'r/2v_r$$
$$- (z - WQr - Wu_B)'(z - WQr - Wu_B)/2v_n$$
$$- u_B'u_B/2v_B + m(q)\ln(\lambda)$$
$$+ [N - m(q)]\ln(1 - \lambda) \qquad (7\text{-}119)$$

where

$$\Omega = v_r WQW' + v_B WW' + v_n I . \qquad (7\text{-}120)$$

In $\mathcal{M}$ the dependence on the wavelet's parameters occurs only in the matrix Ω, through its dependence on the wavelet matrix W. Unfortunately, the second term in $\mathcal{M}$, $z'\Omega^{-1}z/2$, depends on the inverse of Ω, which, as we will see, makes the calculation of the gradients of $\mathcal{M}$ with respect to the wavelet parameters quite complicated. In $\mathcal{L}$, on the other hand, the dependence on the wavelet's parameters occurs just in the third term, $-(z - WQr - Wu_B)'(z - WQr - Wu_B)/2v_n$. We will see that the calculation of the gradients of $\mathcal{L}$ with respect to the wavelet's parameters is quite simple . This is a bit of a paradox. The Separation Principle substantially reduces the number of unknowns from $2n + 4 + 3N$ to $2n + 4 + N$; however, the resulting function $\mathcal{M}$ is exceedingly more difficult to work with for calculations of gradients (and Hessians) than is the original one, $\mathcal{L}$. Whoever said "He giveth with one hand and taketh with the other," or "You never get something for nothing," certainly knew what they were talking about in our situation.

There is also a bit of a problem in calculating the gradients of $\mathcal{L}$ with respect to the wavelet's parameters, but this is not so obvious, and is getting ahead of our story.

7.10.1 Gradients of $\mathcal{M}$ With Respect to **a** and **b**

To begin, we shall compute the gradients of $\mathcal{M}$ with respect to a and b. The following *12 step algorithm* can be used to compute $\partial M / \partial a$ and $\partial M / \partial b$. Each of these steps occurs during the i th iteration; so, all quantities should be subscripted with an i. In order to simplify our already complicated notation, we shall omit this additional notation.

1. *Initialize* a_i *and* b_i. When $i = 0$ the initial values a_0 and b_0 must be provided. How to do this has been briefly discussed in Chapter 4. For all other values of i, a_i and b_i will be available directly from the most recent iteration of the Marquardt-Levenberg algorithm.

2. *Compute the Matrix* $W = W(a_i, b_i)$. This is done by solving the finite-difference equation

$$w(k + n) + a_1 w(k + n - 1) + \ldots + a_{n-1} w(k + 1) + a_n w(k)$$
$$= b_1 \delta(k + n - 1) + b_2 \delta(k + n - 2) + \ldots + b_{n-1} \delta(k + 1)$$
$$+ b_n \delta(k) , \qquad (7\text{-}121)$$

for $w(0), w(1), \ldots, w(N - 1)$, where $w(-n) = w(-n + 1) = \ldots = w(-1) = 0$. In this equation $\delta(k)$ is the unit spike function which has the value of unity at $k = 0$ and is zero at all other values of k. Because the wavelet $w(k)$ is always time limited and stable, its values after some time $k = M$ will be zero; hence, one usually does not have to solve (7-121) all the way up to $k = N - 1$. Equation (7-121) can also be written in operator form as

$$A(z)w(k) = B(z)\delta(k) , \qquad (7\text{-}122)$$

where

$$A(z) = z^n + a_1 z^{n-1} + \ldots + a_{n-1}z + a_n \qquad (7\text{-}123)$$

and

$$B(z) = b_1 z^{n-1} + b_2 z^{n-2} + \ldots + b_{n-1} z + b_n . \qquad (7\text{-}124)$$

This follows from the Chapter 2 fact that

$$W(z) \;= B(z)/A(z) = (b_1 z^{n-1} + b_2 z^{n-2} + \ldots + b^{n-1} z + b_n)/$$
$$(z^n + a_1 z^{n-1} + \ldots + a_{n-1} z + a_n), \qquad (7\text{-}125)$$

and from the fact that a transfer function, such as $W(z)$, is the z-transform of a system's impulse response.

3. *Compute* $\partial w(k)/\partial a_1$ *for* $k = 1, 2, \ldots , N$. This is done by solving the linear finite-difference equation

$$A(z)[\partial w(k)/\partial a_1] = - w(k + n - 1). \qquad (7\text{-}126)$$

Equation (7-126) was obtained by taking the partial derivative of both sides of Eq. (7-121) with respect to the coefficient a_1. Observe that the right-hand side of Equation (7-126) is generated from the results obtained in Step 2. Letting

$$s_{a1}(k) = \partial w(k)/\partial a_1 , \qquad (7\text{-}127)$$

we can express Eq. (7-126) as

$$s_{a1}(k + n) + a_1 s_{a1}(k + n - 1) + \ldots + a_n s_{a1}(k)$$
$$= - w(k + n - 1), \qquad (7\text{-}128)$$

which clearly demonstrates the linear nature of the finite-difference equation for $\partial w(k)/\partial a_1$. We have used the symbol $s_{a1}(k)$ for $\partial w(k)/\partial a_1$ because this partial derivative is often referred to as a *sensitivity function*. Note that, because $w(- n) = w(- n + 1) = \ldots = w(-1) = 0$, $s_{a1}(- n) = s_{a1}(- n + 1) = \ldots = s_{a1}(-1) = 0$.

4. *Compute* $\partial w(k)/\partial a_j$ *for* $j = 2, 3, \ldots , n$ *and* $k = 1, 2, \ldots , N$. This is done by using the fact, which we prove next, that

$$\partial w(k)/\partial a_j = \partial w(k - j + 1)/\partial a_1 \quad \text{for all j and k;} \qquad (7\text{-}129)$$

hence, $\partial w(k)/\partial a_j$ is obtained from $\partial w(k)/\partial a_1$ simply by shifting the latter's arguments and properly storing the associated values, e.g., $\partial w(k)/\partial a_4 = \partial w(k - 3)/\partial a_1 = s_{a1}(k - 3)$.

Proof: Take the partial derivative of both sides of Eq. (7-121) with respect to coefficient a_j to see that

$$A(z)[\partial w(k)/\partial a_j] = -w(k+n-j). \qquad (7\text{-}130)$$

Set $k = k - j + 1$ in Eq. (7-126), to see that

$$A(z)[\partial w(k-j+1)/\partial a_1] = -w(k+n-j). \qquad (7\text{-}131)$$

Compare Eqs. (7-130) and (7-131) to obtain the result in Eq. (7-129).

5. *Compute* $\partial w(k)/\partial b_1$ *for* $k = 1, 2, \dots, N$. This is done by solving the linear finite-difference equation

$$A(z)[\partial w(k)/\partial b_1] = -\delta(k+n-1), \qquad (7\text{-}132)$$

which was obtained by taking the partial derivative of both sides of Eq. (7-121) with respect to b_1. Note that, because $w(-n) = w(-n+1) = \dots = w(-1) = 0$, $s_{b1}(-n) = s_{b1}(-n+1) = \dots = s_{b1}(-1) = 0$.

6. *Compute* $\partial w(k)/\partial b_j$ *for* $j = 2, 3, \dots, n$ *and* $k = 1, 2, \dots, N$. This is done by using the fact that

$$\partial w(k)/\partial b_j = \partial w(k-j+1)/\partial b_1. \qquad (7\text{-}133)$$

The proof of this fact is so similar to the proof of Eq. (7-129) that we leave its details to the reader.

7. *Construct the* $2n$ $N\times N$ *matrices* $\partial W/\partial a_j$ *and* $\partial W/\partial b_j$ *for* $j = 1,2,\dots,n$. Recall, from Chapter 2, that

$$W = \begin{pmatrix} w(0) & 0 & \cdots & 0 \\ w(1) & w(0) & \cdots & 0 \\ \cdot & \cdot & \cdot & \cdot \\ \cdot & \cdot & \cdot & \cdot \\ \cdot & \cdot & \cdot & \cdot \\ w(N-1) & w(N-2) & \cdots & w(0) \end{pmatrix}.$$

Taking the partial derivative of W with respect to a_j, we find that

$$\partial W/\partial a_j = \begin{pmatrix} \partial w(0)/\partial a_j & 0 & \cdots & 0 \\ \partial w(1)/\partial a_j & \partial w(0)/\partial a_j & \cdots & 0 \\ \cdot & \cdot & \cdot & \cdot \\ \cdot & \cdot & \cdot & \cdot \\ \cdot & \cdot & \cdot & \cdot \\ \partial w(N-1)/\partial a_j & \partial w(N-2)/\partial a_j & \cdots & \partial w(0)/\partial a_j \end{pmatrix}. \qquad (7\text{-}134)$$

A similar equation exists for $\partial W/\partial b_j$. Now we see why we need the results obtained in Steps 3, 4, 5, and 6.

8. *Compute $\partial\Omega/\partial a_j$ and $\partial\Omega/\partial b_j$ for* j = 1, 2, ... , n. Observe, from Eq. (7-120), that

$$\begin{aligned}\partial\Omega/\partial a_j &= [\partial W/\partial a_j](v_r Q + v_B I)W' \\ &+ W(v_r Q + v_B I)[\partial W'/\partial a_j]\end{aligned} \qquad (7\text{-}135)$$

and

$$\begin{aligned}\partial\Omega/\partial b_j &= [\partial W/\partial b_j](v_r Q + v_B I)W' \\ &+ W(v_r Q + v_B I)[\partial W'/\partial b_j].\end{aligned} \qquad (7\text{-}136)$$

The results obtained from Steps 2 and 7 are needed to compute $\partial\Omega/\partial a_j$ and $\partial\Omega/\partial b_j$.

9. *Compute matrix Ω.* Recall that

$$\Omega = v_r WQW' + v_B WW' + v_n I . \qquad (7\text{-}137)$$

The results from Step 2 are needed to compute Ω.

10. *Solve the N×N system of linear equations*

$$\Omega y = z \qquad (7\text{-}138)$$

for the N×1 intermediate vector y. Solving Eq. (7-138) is equivalent to determining $\Omega^{-1}z$. Observe that $\mathcal{M}\{a, b, s, q \mid z\}$, given in Eq. (7-118), can also be expressed in terms of the intermediate vector y, as

$$\begin{aligned}\mathcal{M}\{a, b, s, q \mid z\} &= -\tfrac{N}{2}\ln(v_n v_r v_B) - \tfrac{1}{2}y'z + m(q)\ln(\lambda) \\ &+ [N - m(q)]\ln(1 - \lambda).\end{aligned} \qquad (7\text{-}139)$$

For numerical accuracy, orthogonal transformation techniques should be used to solve Eq. (7-138). The LINPACK (Dongarra, et al, 1979) collection of programs is highly recommended.

11. *Compute $\partial y/\partial a_j$ and $\partial y/\partial b_j$ for* j = 1, 2, ..., n. This is accomplished by solving the two sets of linear equations

$$\Omega[\partial y/\partial a_j] = - [\partial\Omega/\partial a_j]y \qquad (7\text{-}140)$$

and

$$\Omega[\partial y/\partial b_j] = - [\partial\Omega/\partial b_j]y. \qquad (7\text{-}141)$$

These equations were obtained by taking the partial derivative of Eq. (7-138) with respect to a_j and b_j, respectively. Observe that results from Steps 8, 9, and 10 are needed in order to compute $\partial y/\partial a_j$ and $\partial y/\partial b_j$. As in Step 10, orthogonal transformation techniques should be used to solve Eqs. (7-140) and (7-141).

12. *Compute $\partial \mathcal{M}\{a, b, s, q \mid z\}/\partial a$ and $\partial \mathcal{M}\{a, b, s, q \mid z\}/\partial b$.* This can now be done, because

$$\partial \mathcal{M}\{a, b, s, q \mid z\}/\partial a = - (\partial y/\partial a_1 \mid \partial y/\partial a_2 \mid$$
$$... \mid \partial y/\partial a_n)'z/2 \qquad (7\text{-}142)$$

and

$$\partial \mathcal{M}\{a, b, s, q \mid z\}/\partial b = - (\partial y/\partial b_1 \mid \partial y/\partial b_2 \mid$$
$$... \mid \partial y/\partial b_n)'z/2 , \qquad (7\text{-}143)$$

which follows from Eq. (7-139).

This completes the 12 step procedure for computing the gradients of $\mathcal{M}$ with respect to **a** and **b**. The procedure requires alot of computational time and storage. Steps 10 and 11 require the solution of $2n + 1$ $N{\times}1$ linear equations. Solving a linear system of N equations requires on the order of N^3 flops (floating point additions and multiplications), although faster methods may be available for our linear system because Ω is a covariance matrix. Consequently, steps 10 and 11 require on the order of $(2n + 1)N^3$ flops. Suppose, for example, it takes 10^{-6} sec to execute 1 flop. Suppose, also, that $n = 6$ and $N = 1,000$. Then the 13×10^9 flops that will be required to solve the 13 $1,000 \times 1,000$ equations will take 13,000 sec (i.e., more than 3 hours), which is a prohibitive amount of CPU time. Of course, by clever coding this time can be reduced, perhaps by a factor of 50%; however, the resulting computation time is still prohibitive, especially since our Block Component Method is an iterative procedure. *In Chapter 9 we shall show that the gradient calculations can be performed, using recursive techniques, in under 3 seconds for this example.* This tremendous savings in computation time is the main reason for using recursive techniques.

Storage of large $N{\times}N$ matrices, such as Ω, can also be a problem. *In Chapter 9 we will also see that the recursive techniques eliminate the need for storing $N{\times}N$ matrices.*

In Chapter 2 we argued that it is much more expedient to use an ARMA model than an MA model. Suppose, however, that we insist on using an MA model, one of length M. Then $\mathcal{M}$ (or $\mathcal{L}$) must be viewed as a function of the M+1 wavelet parameters $w(0), w(1), ..., w(M)$ instead of the 2n ARMA a- and b-parameters. We leave it to the reader to develop a procedure for computing the

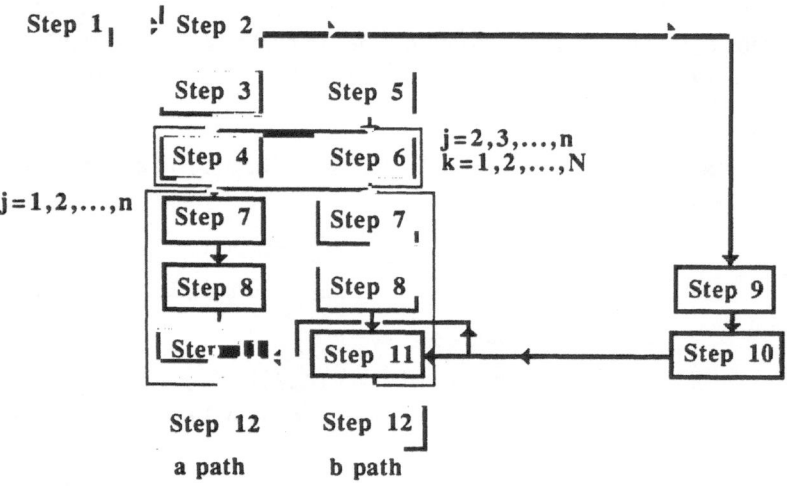

Figure 7-8. Parallel implementation of the 12 Step procedure for computing the gradients of $\mathcal{M}$ with respect to **a** and **b**.

derivatives $\partial\mathcal{M}/\partial w(j)$, $j = 0,1,...,M$; however, what he will discover is that his procedure will contain steps analogous to steps 10 and 11 of our 12 step procedure. Step 11 will now require solution of $M + 1$ $N{\times}1$ linear equations. Because M is usually much larger than 2n, the number of flops needed to execute steps 10 and 11 will be much larger for the MA model than for the ARMA model.

In summary, we must view the batch methods as theoretical, but not as computationally practical.

If parallel processing is possible, then many of the calculations in this 12 Step procedure can be done in parallel. Figure 7-8 depicts the parallelism that is inherent in this 12 Step procedure. A careful count reveals that we need $8 + 6n + 2(n - 1)N$ processors to implement this architecture. The processors need not all be the same. They range from very simple ones, that just perform the shifts needed to compute $\partial w(k)/\partial a_j$ and $\partial w(k)/\partial b_j$ in Steps 4 and 6, to more complicated ones that solve systems of linear equations. Parallelism can also be exploited within each block, using parallel signal processing techniques (e.g., Kung, 1987, and Kung, et al, 1985, 1986); but, this is beyond the scope of this book. We conjecture that at least a 1,000-fold savings in computation time can be achieved using parallel processing. This could make a parallel implementation of the batch formulas computationally competitive with the Chapter 9 recursive formulas; however, the latter can also be parallelized. Consequently, there may be no reason to ever want to compute with the batch formulas.

7.10.2 Gradients of $\mathcal{L}$ With Respect to $\underline{a}$ and $\underline{b}$

Next, we shall compute the gradients of $\mathcal{L}$ with respect to **a** and **b**. The following **9 step algorithm** can be used to compute $\partial\mathcal{L}/\partial\mathbf{a}$ and $\partial\mathcal{L}/\partial\mathbf{b}$. As in the preceding section, each of these steps occurs during the i th iteration; so, all quantities should be subscripted with an i.

Steps 1 through 7 are exactly the same as in the preceding 12 Step procedure. For convenience to the reader, we list these steps:

1. *Initialize* $\mathbf{a}_i$ *and* $\mathbf{b}_i$.

2. *Compute the Matrix* $\mathbf{W} = \mathbf{W}(\mathbf{a}_i, \mathbf{b}_i)$.

3. *Compute* $\partial w(k)/\partial a_1$ *for* $k = 1, 2, \ldots, N$.

4. *Compute* $\partial w(k)/\partial a_j$ *for* $j = 2, 3, \ldots, n$ *and* $k = 1, 2, \ldots, N$.

5. *Compute* $\partial w(k)/\partial b_1$ *for* $k = 1, 2, \ldots, N$.

6. *Compute* $\partial w(k)/\partial b_j$ *for* $j = 2, 3, \ldots, n$ *and* $k = 1, 2, \ldots, N$.

7. *Construct the 2n matrices* $\partial \mathbf{W}/\partial a_j$ *and* $\partial \mathbf{W}/\partial b_j$ *for* $j = 1, 2, \ldots, n$.

8. *Compute* $\partial \mathbf{W}'\mathbf{W}/\partial a_j$ *and* $\partial \mathbf{W}'\mathbf{W}/\partial b_j$ *for* $j = 1, 2, \ldots, n$. This is accomplished using the formulas

$$\partial\mathbf{W}'\mathbf{W}/\partial a_j = [\partial\mathbf{W}'/\partial a_j]\mathbf{W} + \mathbf{W}'[\partial\mathbf{W}/\partial a_j] \qquad (7\text{-}144)$$

and

$$\partial\mathbf{W}'\mathbf{W}/\partial b_j = [\partial\mathbf{W}'/\partial b_j]\mathbf{W} + \mathbf{W}'[\partial\mathbf{W}/\partial b_j]. \qquad (7\text{-}145)$$

Observe that results from Steps 2 and 7 are needed to compute $\partial\mathbf{W}'\mathbf{W}/\partial a_j$ and $\partial\mathbf{W}'\mathbf{W}/\partial b_j$.

9. *Compute* $\partial\mathcal{L}\{a, b, s, q, r, u_B \mid z\}/\partial a$ *and* $\partial\mathcal{L}\{a, b, s, q, r, u_B \mid z\}/\partial b$. This can now be done, because

$$\partial\mathcal{L}\{a, b, s, q, r, u_B \mid z\}/\partial a_j = -z'[\partial\mathbf{W}/\partial a_j]u - u'[\partial\mathbf{W}/\partial a_j]z$$
$$+ u'[\partial\mathbf{W}'\mathbf{W}/\partial a_j]u \qquad (7\text{-}146)$$

and

$$\partial\mathcal{L}\{a, b, s, q, r, u_B \mid z\}/\partial b_j = -z'[\partial\mathbf{W}/\partial b_j]u - u'[\partial\mathbf{W}/\partial b_j]z$$
$$+ u'[\partial\mathbf{W}'\mathbf{W}/\partial b_j]u \qquad (7\text{-}147)$$

for $j = 1, 2, \ldots, n$.   These equations were obtained by taking the partial derivatives of both sides of Eq. (7-119) with respect to a_j and b_j, respectively, and letting

$$\mathbf{u} = \mathbf{Qr} + \mathbf{u_B}. \tag{7-148}$$

Results from Steps 7 and 8 are needed in order to compute $\partial \mathcal{L}/\partial a_j$ and $\partial \mathcal{L}/\partial b_j$. Where do we get $u(k)$ from in order to compute Eqs. (7-146) and (7-147)? *The only choice we have is to approximate $u(k)$ using the current best estimate of it.*   Referring to Figure 4-20, we see that such an estimate can be obtained just after the Adaptive Detection step in our Block Component Search Algorithm. This estimate is $\mathbf{u}^{MV}(N \mid \mathbf{q})$ which is given in Eq. (7-57).   Observe that Fig. 4-20 must be modified in this case to show the block labelled "Update Random Parameters" both before and after the block labelled "Update Wavelet Parameters," because $\mathbf{u}^{MV}(N \mid \mathbf{q})$ is needed to compute the gradients of $\mathcal{L}$. We do not advocate the recalculation of $\mathbf{u}^{MV}(N \mid \mathbf{q})$ for as long as we iterate within the Figure 4-20 block labelled "Update Wavelet Parameters," because there is no guarantee that the loglikelihood function $\mathcal{L}$ will increase if we do this. If it does not increase then we would have to prematurely exit the "Update Wavelet Parameters" step.   Our experience has been that more iterations of the Marquardt-Levenberg Algorithm can be achieved if the input is not reestimated within the wavelet update block.

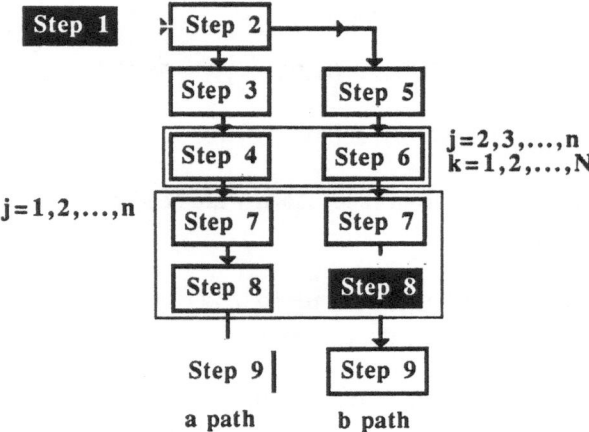

Figure 7-9.   Parallel implementation of the 9 Step procedure for computing the gradients of $\mathcal{L}$ with respect to $\mathbf{a}$ and $\mathbf{b}$.

This completes the 9 Step procedure for computing the gradients of $\mathcal{L}$ with respect to **a** and **b**. *The 9 Step procedure does not require the solution of any* $N \times N$ *equations, which is the reason it is so fast.* It does, however, require computing $\mathbf{u}^{MV}(N \mid q)$ one time.

As was mentioned at the beginning of this section on "Calculating Gradients," "You never get something for nothing." While it may be much faster to calculate the gradients of $\mathcal{L}$ with respect to **a** and **b**, than to calculate the gradients of $\mathcal{M}$ with respect to **a** and **b**, it is less accurate to do so, because we have had to replace the input, which is needed in Step 9 of the 9 Step procedure, with the best available estimate of the input, namely $\mathbf{u}^{MV}(N \mid q)$. No such replacement is necessary in the 12 Step procedure.

If parallel processing is possible, then many of the calculations in the 9 Step procedure can be done in parallel. Figure 7-9 depicts the parallelism that is inherent in this 9 Step procedure. A careful count reveals that we need $6 + 2(n - 1)N + 4n$ processors to implement this architecture. Very large savings in computation time will be achieved by parallel processing.

7.10.3 Derivatives of $\mathcal{M}$ With Respect to Variances

We now compute the derivatives of $\mathcal{M}$ with respect to the two variances v_B and v_n. Recall, in Chapter 4, we explained that, when both the wavelet and the reflectivity are unknown, there is a scale ambiguity which can never be resolved; hence, it is impossible to also estimate the variance of the amplitude sequence, v_r. For this reason we do not bother with the calculations of the derivatives of either $\mathcal{M}$ or $\mathcal{L}$ with respect to v_r. The following *5 Step algorithm* can be used to compute $\partial \mathcal{M}/\partial v_B$ and $\partial \mathcal{M}/\partial v_n$.

1. *Initialize* $v_{B,i}$ *and* $v_{n,i}$. When $i = 0$ the initial values $v_{B,0}$ and $v_{n,0}$ must be provided. Small values are usually chosen. For all other values of i $v_{B,i}$ and $v_{n,i}$ will be available directly from the most recent iteration of the Marquardt-Levenberg algorithm.

2. *Compute the Matrix* $\mathbf{W} = \mathbf{W}(a_i, b_i)$. This is done by using the last values of a_i and b_i that are obtained from the Figure 4-6 (or 4-20) block labelled "Update Wavelet Parameters," and solving finite-difference equation (7-121) for $w(0), w(1), \ldots, w(N - 1)$.

3. *Solve the* $N \times N$ *system of linear equations*

$$\Omega \, y = z \qquad (7\text{-}149)$$

for the $N \times 1$ *intermediate vector* $\mathbf{y}$. The most recent values of $\mathbf{a}_i$ and $\mathbf{b}_i$, obtained from the Figure 4-6 (or 4-20) block labelled "Update Wavelet Parameters," should be used to solve this equation for $\mathbf{y}$. See Step 10 in the previous 12 Step gradient procedure for additional discussions on how to solve Eq. (7-149).

4. *Compute* $\partial \mathbf{y}/\partial v_B$ *and* $\partial \mathbf{y}/\partial v_n$. This is accomplished by solving the two sets of linear equations

$$\Omega[\partial \mathbf{y}/\partial v_B] = \mathbf{WW'y} \qquad (7\text{-}150)$$

and

$$\Omega[\partial \mathbf{y}/\partial v_n] = \mathbf{y} \qquad (7\text{-}151)$$

These equations were obtained by taking the partial derivatives of Eq. (7-149) with respect to v_B and v_n, respectively, using the facts that

$$\partial \Omega/\partial v_B = \mathbf{WW'} \qquad (7\text{-}152)$$

and

$$\partial \Omega/\partial v_n = \mathbf{I} . \qquad (7\text{-}153)$$

Observe that results from Steps 2 and 3 are needed in order to compute $\partial \mathbf{y}/\partial v_B$ and $\partial \mathbf{y}/\partial v_n$.

5. *Compute* $\partial \mathcal{M}\{\mathbf{a}, \mathbf{b}, \mathbf{s}, \mathbf{q} \mid \mathbf{z}\}/\partial v_B$ *and* $\partial \mathcal{M}\{\mathbf{a}, \mathbf{b}, \mathbf{s}, \mathbf{q} \mid \mathbf{z}\}/\partial v_n$. This can now be done, because

$$\partial \mathcal{M}\{\mathbf{a}, \mathbf{b}, \mathbf{s}, \mathbf{q} \mid \mathbf{z}\}/\partial v_B = -(N/2v_B) - [\partial \mathbf{y}/\partial v_B]'\mathbf{z}/2 \qquad (7\text{-}154)$$

and

$$\partial \mathcal{M}\{\mathbf{a}, \mathbf{b}, \mathbf{s}, \mathbf{q} \mid \mathbf{z}\}/\partial v_n = -(N/2v_n) - [\partial \mathbf{y}/\partial v_n]'\mathbf{z}/2 . \qquad (7\text{-}155)$$

These equations were obtained by taking the partial derivative of Eq. (7-139) with respect to v_B and v_n, respectively. Observe that the results from Step 4 are needed in order to compute $\partial \mathcal{M}/\partial v_B$ and $\partial \mathcal{M}/\partial v_n$.

This completes the 5 step procedure for computing the derivatives of $\mathcal{M}$ with respect to v_B and v_n.

Figure 7-10 depicts the parallelism that is inherent in this 5 Step procedure. We need 7 processors to implement this architecture.

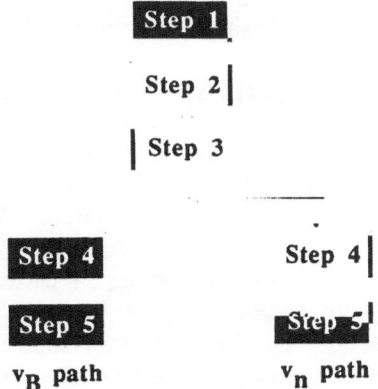

Figure 7-10. Parallel implementation of the 5 Step procedure for computing the derivatives of $\mathcal{M}$ with respect to v_B and v_n.

7.10.4   Derivatives of $\mathcal{L}$ With Respect to Variances

We conclude this section with the calculation of the derivatives of $\mathcal{L}$ with respect to v_B and v_n. These derivatives are computed from the following *4 Step algorithm*.

1. *Initialize $v_{B,i}$ and $v_{n,i}$.* Exactly the same as in Step 1 of the preceding section's 5 Step procedure.

2. *Compute the Matrix $W = W(a_i, b_i)$.* This is done by using the last values of a_i and b_i that are obtained from the Figure 4-6 (or 4-20) block labelled "Update Wavelet Parameters," and solving finite-difference equation (7-121) for $w(0), w(1), \ldots, w(N-1)$.

3. *Compute $u^{MV}(N \mid q)$.* This is done using Eq. (7-57) in which we use the last computed values of a and b, which are available from the Figure 4-20 block labelled "Update Wavelet Parameters," and the values of $v_{B,i}$ and $v_{n,i}$ which are available from Step 1. Observe that Figure 4-20 must be modified in this case to show the block labelled "Update Random Parameters" both before and after the block labelled "Update Wavelet Parameters," because $u^{MV}(N \mid q)$ is needed to compute the gradients of $\mathcal{L}$.

4. *Compute* $\partial\mathcal{L}\{a, b, s, q, r, u_B \mid z\}/\partial v_B$ *and* $\partial\mathcal{L}\{a, b, s, q, r, u_B \mid z\}/\partial v_n$. This is accomplished using the formulas

$$\partial\mathcal{L}\{a, b, s, q, r, u_B \mid z\}/\partial v_B = -(N/2v_B) + u_B'u_B/2v_B{}^2 \quad (7\text{-}156)$$

and

$$\partial\mathcal{L}\{a, b, s, q, r, u_B \mid z\}/\partial v_n = -(N/2v_n)$$
$$+ (z - Wu)'(z - Wu)/2v_n{}^2 . \quad (7\text{-}157)$$

These equations were obtained by taking the partial derivatives of both sides of Eq. (7-119) with respect to v_B and v_n, respectively, and, as in Eq. (7-147), letting $u = Qr + u_B$. In order to compute $\partial\mathcal{L}/\partial v_n$ we must replace u by $u^{MV}(N \mid q)$, which has just been calculated in Step 3. To compute $\partial\mathcal{L}/\partial v_B$ we must also replace u_B by $u_B{}^{MV}(N \mid q)$. Observe, from Eqs. (7-59) and (7-57), that

$$u_B{}^{MV}(N \mid q) = v_B(v_r Q + v_B I)^{-1}u^{MV}(N \mid q) ; \quad (7\text{-}158)$$

hence, it is straightforward to compute $u_B{}^{MV}(N \mid q)$ from $u^{MV}(N \mid q)$.

We leave a block diagram of these simple calculations to the reader.

7.11 Calculating Second Derivatives

In this section we explain how to compute *approximate* second derivative matrices of both $\mathcal{M}$ and $\mathcal{L}$ with respect to the wavelet parameters a and b, and, second derivatives of $\mathcal{M}$ and $\mathcal{L}$ with respect to the two variances v_B and v_n. An approximate second derivative matrix is called a *pseudo-Hessian*. It is used in the Marquardt-Levenberg algorithm in place of an exact Hessian matrix to save on computations.

7.11.1 Pseudo-Hessian of $\mathcal{M}$ With Respect to a and b

Beginning with Eq. (7-139), it is easy show that

$$\partial^2\mathcal{M}/\partial a_i\partial a_j = -[\partial^2 y'/\partial a_i\partial a_j]z/2 , \; i,j = 1, 2, \dots , n \quad (7\text{-}159)$$

and

$$\partial^2 \mathcal{M}/\partial b_i \partial b_j = - [\partial^2 y'/\partial b_i \partial b_j] z/2 \ , \ i,j = 1, 2, \ldots , n \qquad (7\text{-}160)$$

To determine $\partial^2 y/\partial a_i \partial a_j$, take the partial derivative of both sides of Eq. (7-140) with respect to a_i. The result is

$$\Omega [\partial^2 y/\partial a_i \partial a_j] + [\partial \Omega/\partial a_i][\partial y/\partial a_j] = - [\partial^2 \Omega/\partial a_i \partial a_j] y - [\partial \Omega/\partial a_j][\partial y/\partial a_i].$$

The calculations of the second derivatives of Ω are enormously complicated, because they require calculations of $\partial^2 w(k)/\partial a_i \partial a_j$ for $i, j = 1, 2, \ldots , n$ and $k = 1, 2, \ldots , N$. These second-order sensitivity functions are not available from the gradient calculations, and would require solving many finite- difference equations.

Our way around these excessive calculations is to approximate $\partial^2 y/\partial a_i \partial a_j$. This is done simply by deleting the $- [\partial^2 \Omega/\partial a_i \partial a_j] y$ term in this equation; i.e.,

$$\Omega [\partial^2 y/\partial a_i \partial a_j] \simeq - [\partial \Omega/\partial a_i][\partial y/\partial a_j] - [\partial \Omega/\partial a_j][\partial y/\partial a_i] . \qquad (7\text{-}161)$$

where, $i,j = 1, 2, \ldots, n$. Observe that all of the quantities on both sides of this equation have already been computed in Steps 8, 9 and 11 of the 12 Step algorithm for calculating the gradient of $\mathcal{M}$ with respect to $\mathbf{a}$. Equation (7-161) is an $N \times 1$ system of linear equations that must be solved for the $N \times 1$ vector $\partial^2 y/\partial a_i \partial a_j$. A similar equation exists for $\partial^2 y/\partial b_i \partial b_j$. Unfortunately, this procedure requires the solution of $2[n(n - 1)/2]$ $N \times 1$ equations (the Hessian matrix is symmetric), which makes it quite impractical.

Pseudo-Hessians are now used in place of exact Hessians in Eqs. (4-9) and (4-10). Consider $H_{a,i}$. First, $\partial^2 y/\partial a_i \partial a_j$ is solved for from Eq. (7-161). Next, $\partial^2 \mathcal{M}/\partial a_i \partial a_j$ is computed using Eq. (7-159), for $i, j = 1, 2, \ldots , n$. Finally, the $n \times n$ matrix $H_{a,i}$ can be constructed, via Equation (4-12).

7.11.2   Pseudo-Hessian of $\mathcal{L}$ With Respect to $\underline{a}$ and $\underline{b}$

Beginning with Eqs. (7-146) and (7-147), it is easy to show that

$$\partial^2 \mathcal{L}/\partial a_i \partial a_j = - z'[\partial^2 W/\partial a_i \partial a_j] u - u'[\partial^2 W/\partial a_i \partial a_j] z$$
$$+ u'[\partial^2 W'W/\partial a_i \partial a_j] u \qquad (7\text{-}162)$$

and

$$\partial^2 \mathcal{L}/\partial b_i \partial b_j = - z'[\partial^2 W/\partial b_i \partial b_j] u - u'[\partial^2 W/\partial b_i \partial b_j] z$$
$$+ u'[\partial^2 W'W/\partial b_i \partial b_j] u \qquad (7\text{-}163)$$

where, from Eqs. (7-144) and (7-145),

$$\partial^2 W'W/\partial a_i \partial a_j = [\partial^2 W'/\partial a_i \partial a_j]W + [\partial W'/\partial a_i][\partial W/\partial a_j]$$
$$+ W'[\partial^2 W/\partial a_i \partial a_j] + [\partial W'/\partial a_j][\partial W/\partial a_i] \quad (7\text{-}164)$$

and

$$\partial^2 W'W/\partial b_i \partial b_j = [\partial^2 W'/\partial b_i \partial b_j]W + [\partial W'/\partial b_i][\partial W/\partial b_j]$$
$$+ W'[\partial^2 W/\partial b_i \partial b_j] + [\partial W'/\partial b_j][\partial W/\partial b_i] \quad (7\text{-}165)$$

As in the preceding section, we are faced with the computational burden of having to compute $\partial^2 W/\partial a_i \partial a_j$ and $\partial^2 W/\partial b_i \partial b_j$, if we expect to compute the exact Hessian matrices of $\mathcal{L}$ with respect to **a** and **b**. This is too great a burden; hence, we approximate these Hessians by deleting all new second derivative terms in the preceding four equations. The results are (i,j = 1, 2, ..., n):

$$\partial^2 W'W/\partial a_i \partial a_j \simeq [\partial W'/\partial a_i][\partial W/\partial a_j] + [\partial W'/\partial a_j][\partial W/\partial a_i], \quad (7\text{-}166)$$

$$\partial^2 W'W/\partial b_i \partial b_j \simeq [\partial W'/\partial b_i][\partial W/\partial b_j] + [\partial W'/\partial b_j][\partial W/\partial b_i], \quad (7\text{-}167)$$

$$\partial^2 \mathcal{L}/\partial a_i \partial a_j \simeq \mathbf{u}'[\partial^2 W'W/\partial a_i \partial a_j]\mathbf{u}, \quad (7\text{-}168)$$

and

$$\partial^2 \mathcal{L}/\partial b_i \partial b_j \simeq \mathbf{u}'[\partial^2 W'W/\partial b_i \partial b_j]\mathbf{u}, \quad (7\text{-}169)$$

Note that no N×N linear equations must be solved. Of course **u** must be replaced by $\mathbf{u}^{MV}$(N); but, this quantity has already been used in the calculations of $\partial \mathcal{L}/\partial a_i$ and $\partial \mathcal{L}/\partial b_i$, so that it does not have to be recalculated.

7.11.3 Second Derivatives of $\mathcal{M}$ With Respect to Variances

The second derivatives of $\mathcal{M}$ with respect to variances v_B and v_n are remarkably simple. We shall show that

$$\partial^2 \mathcal{M}/\partial v_B^2 = N/(2v_B^2) \quad (7\text{-}170)$$

and

$$\partial^2 \mathcal{M}/\partial v_n^2 = N/(2v_n^2) \quad (7\text{-}171)$$

Proof: Take the derivative of Eq. (7-154) with respect to v_B, and the derivative of Eq. (7-155) with respect to v_n, to show that

$$\partial^2 \mathcal{M}/\partial v_B^2 = N/(2v_B^2) - [\partial^2 \mathbf{y}'/\partial v_B^2]\mathbf{z}/2 \quad (7\text{-}172)$$

and

$$\partial^2 \mathcal{M}/\partial v_n^2 = N/(2v_n^2) - [\partial^2 y'/\partial v_n^2]z/2 \qquad (7\text{-}173)$$

respectively. To compute $\partial^2 y/\partial v_B^2$, take the derivative of Eq. (7-150) with respect to v_B, i.e.,

$$[\partial \Omega/\partial v_B][\partial y/\partial v_B] + \Omega[\partial^2 y/\partial v_B^2] = WW'[\partial y/\partial v_B] \; ;$$

however, from Eq. (7-152), we know that $\partial \Omega/\partial v_B = WW'$; hence, $\Omega[\partial^2 y/\partial v_B^2] = 0$. Because Ω is a covariance matrix it is of full rank, so this last equation means that $[\partial^2 y/\partial v_B^2] = 0$. Consequently, Eq. (7-172) reduces to the simpler result stated in Eq. (7-170). Proceeding in a similar manner, it is easy to show that $[\partial^2 y/\partial v_n^2] = 0$, so that Eq. (7-173) reduces to Eq. (7-171).

7.11.4   Second Derivatives of $\mathcal{L}$ With Respect to Variances

The second derivatives of $\mathcal{L}$ with respect to variances v_B and v_n are also quite simple. Differentiating Eqs. (7-156) and (7-157) with respect to v_B and v_n, respectively, it follows that

$$\partial^2 \mathcal{L}/\partial v_B^2 = N/(2v_B^2) - u'_B u_B/v_B^3 \qquad (7\text{-}174)$$

and

$$\partial^2 \mathcal{L}/\partial v_n^2 = N/(2v_n^2) - (z - Wu)'(z - Wu)/v_n^3 \; . \qquad (7\text{-}175)$$

Of course, MVD estimates of u_B and u must be used in Eqs. (7-174) and (7-175).

7.12 Why v_r Cannot be Estimated: Maximization of $\mathcal{L}$ or $\mathcal{M}$ is an Ill-Posed Problem

In Chapter 4 we pointed out that *when both the wavelet and amplitude variance v_r are unknown, then v_r cannot be determined within a maximum-likelihood framework.* A plausibility argument for this was given. Here we shall give mathematical arguments for it.

Recall that maximizing a loglikelihood function is equivalent to maximizing a likelihood function. Our mathematical arguments are in terms of the functions,

L and M, which are associated with the functions, $\mathcal{L}$ and $\mathcal{M}$, that are given in Eqs. (7-119) and (7-118), respectively. It is easier to present these arguments in terms of L and M. Likelihood function L is given in Eq. (3-9), which we repeat here for the convenience of the reader:

$$
\begin{aligned}
L\{\mathbf{a}, \mathbf{b}, \mathbf{s}, \mathbf{q}, \mathbf{r}, \mathbf{u_B} \mid \mathbf{z}\} = & (2\pi)^{-3N/2}(v_n v_r v_B)^{-N/2}\exp[-\mathbf{r'r}/2v_r \\
& - (\mathbf{z} - \mathbf{WQr} - \mathbf{Wu_B})'(\mathbf{z} - \mathbf{WQr} - \mathbf{Wu_B})/2v_n \\
& - \mathbf{u_B'u_B}/2v_B]\lambda^{m(q)}(1 - \lambda)^{[N - m(q)]}.
\end{aligned} \tag{7-176}
$$

Function M is easily derived from Eq. (7-118) [e.g., compare L and $\mathcal{L}$ in Eqs. (3-10) and (3-9), and $\mathcal{M}$ and $\mathcal{L}$ in Eqs.(7-118) and (7-119)], and is given by the expression

$$
\begin{aligned}
M\{\mathbf{a}, \mathbf{b}, \mathbf{s}, \mathbf{q} \mid \mathbf{z}\} = & (2\pi v_n v_r v_B)^{-N/2}\exp[-\mathbf{z'}\Omega^{-1}\mathbf{z}/2] \\
& \lambda^{m(q)}(1 - \lambda)^{[N - m(q)]}.
\end{aligned} \tag{7-177}
$$

With a little bit of analysis, working directly on Equation (7-176) or (7-177), it is straightforward to show that (note that multiplying $\mathbf{W}$ by a scalar is equivalent to multiplying $\mathbf{b}$ by that scalar)

$$
\begin{aligned}
L\{\mathbf{a}, \mathbf{b}, \mathbf{s}, \mathbf{q}, \mathbf{r}, \mathbf{u_B} \mid \mathbf{z}\} = & L\{\mathbf{a}, \mathbf{b}, v_r, v_B, v_n, \lambda, \mathbf{q}, \mathbf{r}, \mathbf{u_B} \mid \mathbf{z}\} \\
= & L\{\mathbf{a}, \mathbf{b}, v_r^{1/2}, 1, v_B/v_r, v_n, \lambda, \mathbf{q}, \\
& \mathbf{r}/v_r^{1/2}, \mathbf{u_B}/v_r^{1/2} \mid \mathbf{z}\}/(v_r)^N
\end{aligned} \tag{7-178}
$$

and

$$
\begin{aligned}
M\{\mathbf{a}, \mathbf{b}, \mathbf{s}, \mathbf{q} \mid \mathbf{z}\} = & M\{\mathbf{a}, \mathbf{b}, v_r, v_B, v_n, \lambda, \mathbf{q} \mid \mathbf{z}\} \\
= & M\{\mathbf{a}, \mathbf{b}v_r^{1/2}, 1, v_B/v_r, v_n, \lambda, \mathbf{q}, \mid \mathbf{z}\}/(v_r)^N
\end{aligned} \tag{7-179}
$$

These results are due to the convolutional model and the Gaussian assumptions for $\mathbf{r}$ and $\mathbf{u_B}$.

Our mathematical arguments can be made for L or M. They are essentially the same. We present them for L, and our presentation follows that given in Goutsias and Mendel (1986). Observe that, because of Eq. (7-178),

$$
\begin{aligned}
L\{\mathbf{a}, \mathbf{b}, \mathbf{s}, \mathbf{q}, \mathbf{r}, \mathbf{u_B} \mid \mathbf{z}\} = & G_L\{\mathbf{a}, \mathbf{b}v_r^{1/2}, v_B/v_r, v_n, \lambda, \mathbf{q}, \\
& \mathbf{r}/v_r^{1/2}, \mathbf{u_B}/v_r^{1/2} \mid \mathbf{z}\}/(v_r)^N
\end{aligned} \tag{7-180}
$$

To simplify notation we introduce the following generic symbols: $x_1 = \mathbf{a}$, $x_2 = \mathbf{b}$, $x_3 = v_r$, $x_4 = v_B$, $x_5 = v_n$, $x_6 = \lambda$, $x_7 = \mathbf{q}$, $x_8 = \mathbf{r}$, and $x_9 = \mathbf{u_B}$. Consequently, Eq. (7-180) is of the form

$$
\begin{aligned}
L(x_1, x_2, x_3, x_4, x_5, x_6, x_7, x_8, x_9) = & G_L(x_1, x_2 x_3^{1/2}, \\
& x_4/x_3, x_5, x_6, x_7, \\
& x_8/x_3^{1/2}, x_9/x_3^{1/2})/x_3^N,
\end{aligned} \tag{7-181}
$$

which can also be written, as

$$L(x_1, x_2, x_3, x_4, x_5, x_6, x_7, x_8, x_9) = G_L(m_1, m_2, m_3,$$
$$m_4, m_5, m_6,$$
$$m_7, m_8)/x_3^N, \qquad (7\text{-}182)$$

where $m_1 = x_1$, $m_2 = x_2 x_3^{1/2}$, $m_3 = x_4/x_3$, $m_4 = x_5$, $m_5 = x_6$, $m_6 = x_7$, $m_7 = x_8/x_3^{1/2}$, and $m_8 = x_9/x_3^{1/2}$. From Eq. (7-182) and these relationships, it follows that

$$\partial L/\partial x_1 = [\partial L/\partial m_1][\partial m_1/\partial x_1] = [\partial G_L/\partial m_1]/x_3^N,$$

$$\partial L/\partial x_2 = \{[\partial G_L/\partial m_2]/x_3^N\}x_3^{1/2},$$

$$\partial L/\partial x_3 = \partial(G_L/x_3^N)/\partial x_3 + [\partial L/\partial m_2][\partial m_2/\partial x_3] + [\partial L/\partial m_3][\partial m_3/\partial x_3]$$
$$+ [\partial L/\partial m_7][\partial m_7/\partial x_3] + [\partial L/\partial m_8][\partial m_8/\partial x_3],$$
$$= -G_L N x_3^{-(N+1)} + \{[\partial G_L/\partial m_3]/x_3^N\}[-x_4/x_3^2]$$
$$+ \{[\partial G_L/\partial m_2]/x_3^N\}[x_2/2x_3^{1/2}] - \{[\partial G_L/\partial m_7]/x_3^N\}[x_8 x_3^{-3/2}]/2$$
$$- \{[\partial G_L/\partial m_8]/x_3^N\}[x_9 x_3^{-3/2}]/2,$$

$$\partial L/\partial x_4 = \{[\partial G_L/\partial m_3]/x_3^N\}[1/x_3],$$

$$\partial L/\partial x_5 = [\partial G_L/\partial m_4]/x_3^N,$$

$$\partial L/\partial x_6 = [\partial G_L/\partial m_5]/x_3^N,$$

$$\partial L/\partial x_7 = [\partial G_L/\partial m_6]/x_3^N,$$

$$\partial L/\partial x_8 = \{[\partial G_L/\partial m_7]/x_3^N\}[1/x_3^{1/2}],$$

and

$$\partial L/\partial x_9 = \{[\partial G_L/\partial m_8]/x_3^N\}[1/x_3^{1/2}].$$

Observe that if $x_3 \neq 0$ and $G_L \neq 0$, then the nine partial derivatives of L cannot be simultaneously set equal to zero. Indeed, if $\partial L/\partial x_i = 0$ for $i = 1, 2, 4, 5, \ldots,$ 9, then $\partial G_L/\partial m_i = 0$ for $i = 1, 2, \ldots, 8$. From the equation $\partial L/\partial x_3 = 0$, we see that

$$\partial L/\partial x_3 = -G_L N x_3^{-(N+1)} \neq 0;$$

hence, likelihood function L cannot be maximized with respect to all of its parameters. It can only be maximized with respect to $x_1, x_2, x_4, x_5, x_6, x_7,$

x_8, and x_9. Although an optimum point $(x_1^*, x_2^*, x_3^*, x_4^*, x_5^*, x_6^*, x_7^*, x_8^*, x_9^*)$ of L does not exist, an optimum point $(m_1^*, m_2^*, m_3^*, m_4^*, m_5^*, m_6^*, m_7^*, m_8^*)$ of function G_L may exist.

Consequently, if the true value of v_r is unknown, the function G_L must be maximized instead of the function L. According to Eq.(7-180), G_L can be easily obtained from L by setting $v_r = 1$. In this case, estimates b^*, v_B^*, r^*, and u_B^* of b, v_B, r, and u_B can only be obtained to within an unknown scale factor. *In effect maximization of L (or M) is an ill-posed problem. By setting v_r equal to different values, different solutions will be obtained for maximization of L (or M).* As we show in Chapter 8, no such problem occurs for the maximization of P (or N).

7.13 An Algorithm for λ

When we examine the two functions $\mathcal{M}\{a, b, s, q \mid z\}$ and $\mathcal{L}\{a, b, s, q, r, u_B \mid z\}$, in Eqs. (7-118) and (7-119), respectively, we see that they both depend on the statistical parameter λ in exactly the same way. Only their last two terms --- $m(q)\ln(\lambda) + [N - m(q)]\ln(1 - \lambda)$ --- depend upon λ. Consider $\mathcal{M}$ for example, i.e.,

$$\mathcal{M}\{a, b, s, q \mid z\} = -\tfrac{N}{2}\ln(v_n v_r v_B) - \tfrac{1}{2}z'\Omega^{-1}z \\ + m(q)\ln(\lambda) + [N - m(q)]\ln(1 - \lambda). \qquad (7\text{-}183)$$

The maximum-likelihood value of λ is obtained by setting the partial derivative of $\mathcal{M}$ equal to zero:

$$\partial \mathcal{M}/\partial \lambda = m(q)(1/\lambda) - [N - m(q)][1/(1 - \lambda)] = 0.$$

There are three solutions to this equation, namely 0, 1, and $m(q)/N$. By substituting each of these solutions into Eq. (7-183), it is straightforward to show that the solutions 0 and 1 minimize $\mathcal{M}$, whereas $m(q)/N$ maximizes $\mathcal{M}$; hence,

$$\lambda^{ML} = m(q)/N. \qquad (7\text{-}184)$$

At this point there are two possibilities: (1) event sequence q is known, in which case $m(q) = m$, where m is the number of *known* time points at which a unity appears in the event sequence; and (2) event sequence q is unknown, in which case how can we use Eq. (7-184) to evaluate λ^{ML}?

Recall that, in the second case, we use a detector to determine q. However, all of our detectors require knowledge of λ in order to implement their decision strategies. Obviously, detection and determining λ are coupled problems; hence, the adaptive detection block in Figure 4-20. Suppose we are given the i th value of λ, λ_i. This value of λ is what we use in any one of our detectors to obtain q_{DETCT}. Note that $q_{DETCT} = q_{DETCT}(\lambda_i)$. We then use $q_{DETCT}(\lambda_i)$ to compute λ_{i+1}, i.e.,

$$\lambda_{i+1} = m[q_{DETCT}(\lambda_i)]/N. \tag{7-185}$$

This is the relaxation algorithm given in Eq. (4-22).

These same results are also obtained when we begin with the loglikelihood function $\mathcal{N}\{a, b, s, q \mid z\}$, where $\mathcal{N}$ is defined in Equation (5-11).

8
Mathematical Details for Chapter 5

8.1 Introduction

This chapter is similar in spirit to Chapter 7. It provides the mathematical details for many of the statements that were made, without proof, in Chapter 5. For convenience of the reader, we again state many of the Chapter 5 results, prior to their derivations.

8.2 MVD Filter Properties

In this section we first show that the Fourier transform, $F(\omega)$, of a constant coefficient doubly-infinite two-sided MVD filter is given by

$$F(\omega) = v_u W^*(\omega)/[v_u |W(\omega)|^2 + v_n], \qquad (8\text{-}1)$$

and, then, that the convolution of the resolution function, $\rho(k)$, and reflectivity, $u(k)$, tends to undershoot $u(k)$.

8.2.1 Derivation of $F(\omega)$

The situation is that of Figure 5-2 in which $d(k) = u(k)$. We choose the two-sided deconvolution filter's coefficients to minimize the mean-squared error, $I(f)$, between the filter's output, $y(k) = f(k)*z(k)$, and the desired signal, $u(k)$, i.e., we choose $f = \text{col}[f(0), f(\pm 1), f(\pm 2), ...]$ to minimize

$$I(f) = E\left\{ \sum_{k=0}^{\infty} [u(k) - \sum_{m=-\infty}^{\infty} f(m)z(k - m)]^2 \right\}, \qquad (8\text{-}2)$$

Taking the partial derivative of I(f) with respect to the j th filter coefficient, f(j), interchanging the order of expectation and summation, and setting the result equal to zero, we find, that

$$\sum_{m=-\infty}^{\infty} f(m)\phi_{zz}(j-m) = \phi_{zu}(m) \quad \text{where } j = 0, \pm 1, \pm 2, \ldots \tag{8-3}$$

We are now ready to solve (8-3) for the deconvolution filter's coefficients. We cannot do this by solving a linear system of equations because there are a doubly infinite number of them. Instead, we take the discrete-time Fourier transform of (8-3), i.e.,

$$F(\omega)\Phi_{zz}(\omega) = \Phi_{zu}(\omega) \tag{8-4}$$

We must now determine $\Phi_{zu}(\omega)$ and $\Phi_{zz}(\omega)$. To begin, we first evaluate $\phi_{zu}(m)$ and $\phi_{zz}(j)$.

In order to evaluate $\phi_{zu}(m)$ we use the convolutional model for z(k) given in Eq. (2-12), i.e. (Mendel, 1987),

$$\phi_{zu}(m) = E\{z(k-m)u(k)\} = \sum_{i=1}^{k} E\{u(i)u(k)\}w(k-m-i) + E\{n(k-m)u(k)\}$$

$$= v_u w(-m) \tag{8-5}$$

because $E\{u(i)u(k)\} = v_u\delta(i-k)$ and $E\{n(k-m)u(k)\} = E\{n(k-m)\}E\{u(k)\} = 0$.

In order to evaluate $\phi_{zz}(j)$, we again begin with the convolutional model, but, so as not to become encumbered with awkward limits on summations and associated transformations of variables, we embed our convolutional model

$$z(k) = \sum_{i=1}^{k} u(i)w(k-i) + n(k) \quad k = 1, 2, \ldots, N \tag{8-6}$$

into the two-sided model

$$z(k) = \sum_{i=-\infty}^{\infty} u(i)w(k-i) + n(k) \quad k = 1, 2, \ldots, N \tag{8-7}$$

For causal systems and inputs that don't start until time zero, the second form does indeed reduce to the first form. We also use the facts that u(k) and n(k) are stationary, white, uncorrelated, and zero mean; hence,

$$\phi_{zz}(j) = E\{z(k)z(k+j)\} = E\{\sum_{i,l=-\infty}^{\infty}\sum u(i)u(l)w(k-i)w(k-l+j)\}$$

$$+ E\{n(k)n(k+j)\}$$

$$= \sum_{i,l=-\infty}^{\infty}\sum v_u \delta(i-l)w(k-i)w(k-l+j) + v_n \delta(j)$$

$$= v_u \sum_{i=-\infty}^{\infty} w(k-i)w(k-i+j) + v_n \delta(j)$$

$$= v_u \sum_{n=-\infty}^{\infty} w(n)w(n+j) + v_n \delta(j) \tag{8-8}$$

which is the desired result.

From (8-5), we find that

$$\Phi_{zu}(\omega) = \mathcal{F}\{\phi_{zu}(m)\} = \mathcal{F}\{v_u w(-m)\} = v_u W^*(\omega) \tag{8-9}$$

and, from (8-8), we find that

$$\Phi_{zz}(\omega) = \mathcal{F}\{\phi_{zz}(m)\} = \mathcal{F}\{[v_u \sum_{n=-\infty}^{\infty} w(n)w(n+j) + v_n \delta(j)]\}$$

$$= v_u W(\omega)W(-\omega) + v_n ;$$

but, $W(-\omega)$ is the complex conjugate of $W(\omega)$, so that

$$\Phi_{zz}(\omega) = v_u \mid W(\omega) \mid^2 + v_n \tag{8-10}$$

Substitute (8-9) and (8-10) into (8-4), to obtain the desired result in (8-1).

8.2.2 Undershoot Property

Assume that (Chi and Mendel, 1984) sequence $u(k)$ contains only an *isolated spike* at $k = k_1$. Let $u_s^{MV}(k \mid N)$ denote the 'signal' component in $u^{MV}(k \mid N)$, where, from Eq. (5-5), $u_s^{MV}(k \mid N) = \rho(k) * u(k)$, in which $\rho(k)$ is the resolution function whose discrete-time Fourier transform is given in Eq. (5-7). Then, we shall show that

$$\mid u_s^{MV}(k \mid N) \mid \leqslant \mid u_s^{MV}(k_1 \mid N) \mid < \mid u(k_1) \mid \tag{8-11}$$

which means that $u_s^{MV}(k_1 \mid N)$ undershoots $u(k_1)$.

Proof: Sequence $u(k)$ can be expressed as $u(k) = u(k_1)\delta(k - k_1)$ so that

$$u_s^{MV}(k \mid N) = u(k_1)\rho(k - k_1) \qquad (8\text{-}12)$$

Because $\rho(k)$ is a zero phase function it behaves just like an autocorrelation function. A basic property of an autocorrelation function $\rho(k)$ is that $\rho(0) \geqslant \mid \rho(j) \mid$ for all $j \neq 0$. The maximum value of $\rho(k - k_1)$ therefore occurs at $k = k_1$; hence,

$$\mid u_s^{MV}(k \mid N) \mid \leqslant \mid u_s^{MV}(k_1 \mid N) \mid = \rho(0) \mid u(k_1) \mid \qquad (8\text{-}13)$$

We must now show that $\rho(0) < 1$. From Eq. (5-7), we see that

$$0 \leqslant R(\omega) < 1 \text{ for } \forall -\pi \leqslant \omega \leqslant \pi \qquad (8\text{-}14)$$

From Parseval's relation (e.g., Papoulis, 1984) and Eq. (8-14), we determine that

$$\sum_{k=-\infty}^{\infty} \rho^2(k) = \frac{1}{2\pi} \int_{-\pi}^{\pi} R^2(\omega)d\omega < \frac{1}{2\pi} \int_{-\pi}^{\pi} d\omega = 1 \qquad (8\text{-}15)$$

from which it follows that $\rho(0) < 1$. Consequently, $\mid u_s^{MV}(k \mid N) \mid \leqslant \mid u_s^{MV}(k_1 \mid N) \mid < \mid u(k_1) \mid$, which was to be proved. $\square$

If $u(k)$ consists of widely separated spikes at $k = k_1, k_2, \ldots, k_M$, then this undershoot property explains why $u_s^{MV}(k \mid N)$ will undershoot $u(k)$ at $k = k_1, k_2, \ldots, k_M$. On the other hand, if $u(k)$ contains closely spaced spikes, then it is quite possible for an overshoot to occur. Additionally, if a lot of noise is present then $u^{MV}(k \mid N)$ may look quite different from $u_s^{MV}(k \mid N)$, and it will be difficult to observe the undershoot property. On the other hand, if not much noise is present, then the undershoot property derived for $u_s^{MV}(k \mid N)$ will also be discernable on $u^{MV}(k \mid N)$.

Example 8-1. This example, which is taken from Chi and Mendel (1984), demonstrates a case in which overshoot occurs between $u^{MV}(k \mid N)$ and $u(k)$. The wavelet is the narrow-band $w_2(k)$, whose transfer function is given in Eq. (6-2). Signal-to-noise ratio equals 50. By design, there are only two close spikes in $u(k)$. Figure 8-1 depicts $u_s^{MV}(k \mid N)$; it undershoots $u(k)$ for the larger spike and overshoots $u(k)$ for the smaller spike. The same situation occurs for $u^{MV}(k \mid N)$, depicted in Figure 8-2. The overshoot behavior depicted in these two figures is due to the close spacing of the two spikes and the broad (in the time-domain) nature of $\rho(k)$ (see Figure 6-22).

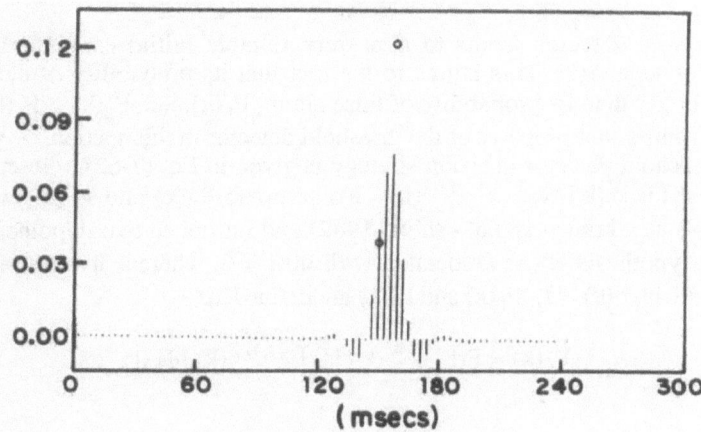

Figure 8-1. $u_s^{MV}(k \mid N)$ (SNR = 50). Circles mark true values of u(k) and bars mark the corresponding estimates. The larger true spike is scaled by a factor of 0.5. (Chi and Mendel, 1984b. © 1984 IEEE.)

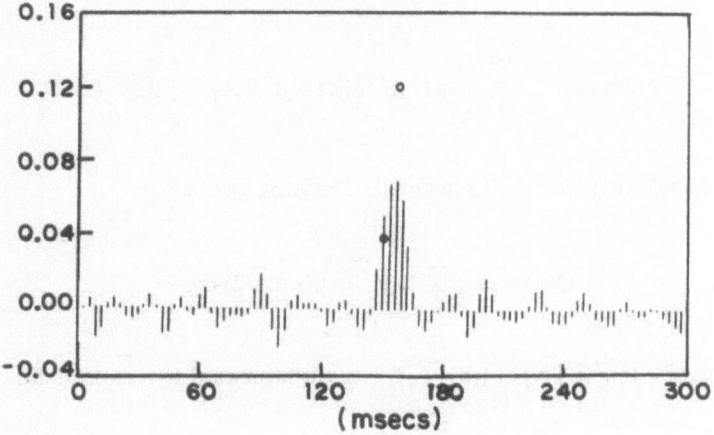

Figure 8-2. $u^{MV}(k \mid N)$ (SNR = 50). Circles mark true values of u(k) and bars mark the corresponding estimates. The larger true spike is scaled by a factor of 0.5. (Chi and Mendel, 1984b. © 1984 IEEE.)

8.3   Threshold Detector

The threshold detector seems to give very reliable initial estimates for the reflectivity sequence. This is due to the fact that its probability of detection, $P_d(k)$, is larger than its probability of false alarm, $P_f(k)$, i.e., $P_d(k) > P_f(k)$. We prove this important property of the threshold detector in this section.

The threshold detector decision strategy is given in Eq. (7-62). Observe that $q(k) = 1$ if $| u^{MV}(k \mid N) | > t^{1/2}(k)$. To compute $P_d(k)$ and $P_f(k)$, we treat $u^{MV}(k \mid N)$ as a known signal (Shiva, 1982) and introduce two hypotheses, H_0 and H_1. Hypothesis H_0 is associated with $q(k) = 0$, whereas hypothesis H_1 is associated with $q(k) = 1$. $P_f(k)$ and $P_d(k)$ are defined, as

$$P_f(k) = \Pr[| u^{MV}(k \mid N) | > t^{1/2}(k) \mid H_0] \tag{8-16}$$

and

$$P_d(k) = \Pr[| u^{MV}(k \mid N) | > t^{1/2}(k) \mid H_1] \tag{8-17}$$

These probabilities can be evaluated, using Eq. (7-73), as

$$P_f(k) = 2[2\pi A_0(k)]^{-1/2} \int_{t^{1/2}(k)}^{\infty} \exp\{- [u^{MV}(k \mid N)]^2/2A_0(k)\}\, du^{MV}(k \mid N) \tag{8-18}$$

and

$$P_d(k) = 2[2\pi A_1(k)]^{-1/2} \int_{t^{1/2}(k)}^{\infty} \exp\{- [u^{MV}(k \mid N)]^2/2A_1(k)\}\, du^{MV}(k \mid N) \tag{8-19}$$

We shall now evaluate $P_f(k)$ in detail. Defining $Q(x)$ as

$$Q(x) = (1/\sqrt{2\pi}) \int_{x}^{\infty} \exp(- \tau^2/2)\, d\tau \tag{8-20}$$

and letting $x = u^{MV}(k \mid N)/A_0(k)^{1/2}$, it is straightforward to reexpress Eq. (8-18), as

$$P_f(k) = 2Q[t^{1/2}(k)/A_0(k)^{1/2}] \tag{8-21}$$

Proceeding similarly for $P_d(k)$, we find

$$P_d(k) = 2Q[t^{1/2}(k)/A_1(k)^{1/2}] \qquad (8\text{-}22)$$

We must now compare $P_d(k)$ and $P_f(k)$. First, observe from Eq. (8-20), that, because the integrand of $Q(x)$ is always positive within the limits of integration, $Q(x) > Q(y)$ when $x < y$. Next, using Eq. (7-64), for $q = 0$ and $q = 1$, it is easy to show that $A_1(k) > A_0(k)$; hence, $Q[t^{1/2}(k)/A_1(k)^{1/2}] > Q[t^{1/2}(k)/A_0(k)^{1/2}]$. Applying these facts to a comparison of $P_d(k)$ and $P_f(k)$ in Eqs. (8-21) and (8-22), it is now obvious that

$$P_d(k) > P_f(k) \qquad (8\text{-}23)$$

8.4 Modified Likelihood Function

To obtain the modified likelihood function given in Eq. (5-10), we start with the original likelihood function given in Eq. (3-2), which we repeat here for the convenience of the reader:

$$L\{a, b, s, q, r, u_B \mid z\} = p(z \mid q, r, u_B, \\ a, b, s) \, Pr(q \mid s) \, p(r \mid s) \, p(u_B \mid s) \qquad (8\text{-}24)$$

If we integrate both sides of Eq. (8-24) with respect to r and u_B, using

$$p(z \mid q, a, b, s) = \int_{-\infty}^{\infty} \cdots \int_{-\infty}^{\infty} p(z \mid q, r, u_B, a, b, s) \, p(r \mid s) \, p(u_B \mid s) \, dr \, du_B , \quad (8\text{-}25)$$

we can determine $q, a, b,$ and s using the 'modified' likelihood expression

$$N\{a, b, s, q \mid z\} = p(z \mid q, a, b, s) \, Pr(q \mid s) \qquad (8\text{-}26)$$

When q is given, z is Gaussian, so that

$$p(z \mid q, a, b, s) = (2\pi)^{-N/2} \mid \Omega \mid^{-1/2} \exp(- z'\Omega^{-1}z/2) \qquad (8\text{-}27)$$

$Pr(q \mid s)$ is given in Eq. (3-5). Combining Eqs. (8-27), (3-5) and (8-26), we obtain $N\{a, b, s, q \mid z\}$ given in Eq. (5-10), i.e.,

$$N\{a, b, s, q \mid z\} = (2\pi)^{-N/2} \mid \Omega \mid^{-1/2} \exp(- z'\Omega^{-1}z/2) \\ \lambda^{m(q)} (1 - \lambda)^{[N - m(q)]} \qquad (5\text{-}10)$$

8.5   Separation Principle for P and Derivation of N From P

Here we show that the Separation Principle, which is associated with L (or $\mathcal{L}$) applies as well to P (or $\mathcal{P}$), where [e.g., see Eq. (5-27)]

$$P\{a, b, s, q, r, u_B \mid z\} \triangleq$$
$$(2\pi)^N (v_n v_r v_B)^{N/2} \mid \Omega \mid^{-1/2} L\{a, b, s, q, r, u_B \mid z\} \qquad (8\text{-}28)$$

and, that after the Separation Principle is applied to P, we obtain the likelihood function N, given in Eq. (5-10).

Applying step 1 of the Mathematical Fact (that is given on the first page of Chapter 7) to function P, with $x \triangleq \text{col}(r, u_B)$, is equivalent to applying step 1 of this Mathematical Fact to function L, because multiplicative factor $(2\pi)^N (v_n v_r v_B)^{N/2} \mid \Omega \mid^{-1/2}$ is independent of x and is nonnegative. Therefore, $r = r^*$ and $u_B = u_B^*$ will still be given by Eqs. (7-15) and (7-14), for any vectors a, b, s, and q.

Substituting the $r = r^*$ and $u_B = u_B^*$ values of r and u_B into P yields, as we show next, N, given in Eq. (5-10). From the Separation Principle, applied to L, we find that [see, for example, Figure 5-3, and Eq. (7-21)]

$$P\{a, b, s, q, r = r^*, u_B = u_B^* \mid z\} = (2\pi)^N (v_n v_r v_B)^{N/2} \mid \Omega \mid^{-1/2}$$
$$L\{a, b, s, q, r = r^*, u_B = u_B^* \mid z\}$$
$$= (2\pi)^N (v_n v_r v_B)^{N/2} \mid \Omega \mid^{-1/2} M\{a, b, s, q \mid z\} \qquad (8\text{-}29)$$

Substituting Eq. (5-23) into Eq. (8-29), we find

$$P\{a, b, s, q, r = r^*, u_B = u_B^* \mid z\} = (2\pi)^{-N/2} \mid \Omega \mid^{-1/2} \exp(-z'\Omega^{-1}z/2)$$
$$\lambda^{m(q)}(1 - \lambda)^{[N - m(q)]}$$
$$= N\{a, b, s, q \mid z\} \qquad (8\text{-}30)$$

This last equality is obtained by comparing the first line of Eq. (8-30) with Eq. (5-25).

8.6 Why v_r Cannot be Estimated: Maximization of P or N is Not an Ill-Posed Problem

The details in this section are quite similar to those in the Chapter 7 section, entitled "Why v_r Cannot be Estimated: Maximization of L is an Ill-Posed

Problem." Here we shall show that when we work with the objective function P or the likelihood function N, v_r still cannot be estimated; but, maximizing P or N is not an ill-posed problem. We shall only provide the details for P.

Beginning with Eqs. (5-27) and (5-21), and some analysis, it is straightforward to show that

$$
\begin{aligned}
P\{\mathbf{a}, \mathbf{b}, \mathbf{s}, \mathbf{q}, \mathbf{r}, \mathbf{u}_B \mid \mathbf{z}\} &= P\{\mathbf{a}, \mathbf{b}, v_r, v_B, v_n, \lambda, \mathbf{q}, \mathbf{r}, \mathbf{u}_B \mid \mathbf{z}\} \\
&= P\{\mathbf{a}, \mathbf{b}v_r^{1/2}, 1, v_B/v_r, v_n, \lambda, \mathbf{q}, \mathbf{r}/v_r^{1/2}, \\
&\quad \mathbf{u}_B/v_r^{1/2} \mid \mathbf{z}\} \\
&= G_P\{\mathbf{a}, \mathbf{b}v_r^{1/2}, v_B/v_r, v_n, \lambda, \mathbf{q}, \mathbf{r}/v_r^{1/2}, \\
&\quad \mathbf{u}_B/v_r^{1/2} \mid \mathbf{z}\}
\end{aligned}
\tag{8-31}
$$

As in Chapter 7, we now let $x_1 = \mathbf{a}$, $x_2 = \mathbf{b}$, $x_3 = v_r$, $x_4 = v_B$, $x_5 = v_n$, $x_6 = \lambda$, $x_7 = \mathbf{q}$, $x_8 = \mathbf{r}$, and $x_9 = \mathbf{u}_B$. Consequently, Eq. (8-31) is of the form

$$
\begin{aligned}
&P(x_1, x_2, x_3, x_4, x_5, \\
&\quad x_6, x_7, x_8, x_9) = G_P(x_1, x_2\, x_3^{1/2}, x_4/x_3, x_5, x_6, x_7, \\
&\quad x_8/x_3^{1/2}, x_9/x_3^{1/2}),
\end{aligned}
\tag{8-32}
$$

which can also be written, as

$$
\begin{aligned}
&P(x_1, x_2, x_3, x_4, x_5, x_6, x_7, \\
&\quad x_8, x_9) = G_P(m_1, m_2, m_3, m_4, m_5, m_6, m_7, m_8),
\end{aligned}
\tag{8-33}
$$

where $m_1 = x_1$, $m_2 = x_2\, x_3^{1/2}$, $m_3 = x_4/x_3$, $m_4 = x_5$, $m_5 = x_6$, $m_6 = x_7$, $m_7 = x_8/x_3^{1/2}$, and $m_8 = x_9/x_3^{1/2}$. From Eq. (8-33) and these relationships, it follows that

$$
\partial P/\partial x_1 = [\partial P/\partial m_1][\partial m_1/\partial x_1] = \partial G_P/\partial m_1 \, ,
$$

$$
\partial P/\partial x_2 = [\partial G_P/\partial m_2]x_3^{1/2} \, ,
$$

$$
\begin{aligned}
\partial P/\partial x_3 &= [\partial P/\partial m_2][\partial m_2/\partial x_3] + [\partial P/\partial m_3][\partial m_3/\partial x_3] \\
&\quad + [\partial P/\partial m_7][\partial m_7/\partial x_3] + [\partial P/\partial m_8][\partial m_8/\partial x_3] \, , \\
&= [\partial G_P/\partial m_2][x_2 x_3^{-1/2}]/2 + [\partial G_P/\partial m_3][-x_4 x_3^{-2}] \\
&\quad - [\partial G_P/\partial m_7][x_8 x_3^{-3/2}]/2 - [\partial G_P/\partial m_8][x_9 x_3^{-3/2}]/2 \, ,
\end{aligned}
$$

$$
\partial P/\partial x_4 = [\partial G_P/\partial m_3]x_3^{-1} \, ,
$$

$$
\partial P/\partial x_5 = \partial G_P/\partial m_4 \, ,
$$

$$
\partial P/\partial x_6 = \partial G_P/\partial m_5 \, ,
$$

$$
\partial P/\partial x_7 = \partial G_P/\partial m_6 \, ,
$$

$$\partial P/\partial x_8 = [\partial G_P/\partial m_7]x_3^{-1/2} \, ,$$

and

$$\partial P/\partial x_9 = [\partial G_P/\partial m_8]x_3^{-1/2}.$$

Observe from these equations that, in order to maximize function P, it suffices to maximize function G_P, i.e., if $\partial G_P/\partial m_i = 0$ for $i = 1, 2, \ldots , 8$, then $\partial P/\partial x_j = 0$ for $j = 1, 2, \ldots , 9$. If $(m_1^*, m_2^*, \ldots , m_8^*)$ is the optimum point of G_P, then any point $(x_1^*, x_2^*, \ldots , x_9^*)$, with $x_1^* = m_1^*$, $x_2^* = m_2^*$ $(x_3^*)^{-1/2}$, $x_4^* = m_3^* x_3^*$, $x_5^* = m_4^*$, $x_6^* = m_5^*$, $x_7^* = m_6^*$, $x_8^* = m_7^* x_3^{*1/2}$, and $x_9^* = m_8^* x_3^{*1/2}$ maximizes function P.

This demonstrates that maximization of P is not an ill-posed problem; however, maximization of P cannot resolve the true value of $x_3 = v_r$. A value for v_r must be set ahead of time. Unlike the ill-posed situation for L, that was described in Chapter 7, here, when we set v_r equal to different values, the *same* solutions will be obtained for the other m_i variables, when P (or N) is maximized.

8.7 SMLR[1] Detector Based on N

A slightly different SMLR detector is obtained when likelihood function N is used instead of function M. The derivation of this SMLR detector follows the steps given in the Chapter 7 section entitled "Single Most-Likely Replacement Detector," simply by replacing M by N. Of course, the explicit formulas for M and N are different [e.g., see (5-23) and (5-25)], so the resulting expression for $\text{Ln}D(z;k)$ will be different. The resulting single most-likely replacement detector is designated an SMLR[1] detector, to distinguish it from the SMLR detector that is based on M.

Let $D^1(z;k)$ denote the SMLR[1] decision function. The SMLR[1] detector decision strategy is to examine all the values of k for which $\text{Ln}D^1(z;k) > 0$ and to find the value of k at which $\text{Ln}D^1(z;k)$ is a maximum. This time point, which we designate k', is then the single time point at which a change is made in our reference sequence. Now

$$\text{Ln}D^1(z;k) = \text{Ln N}\{q_{t,k} \mid z\} - \text{Ln N}\{q_r \mid z\} \tag{8-34}$$

Substituting Eq. (5-25) into (8-34), we obtain the following final expression for $\text{Ln}D^1(z;k)$:

$$
\begin{aligned}
\mathrm{L}n D^1(z;k) &= (w_k{}'P_z^{-1}z)^2/\{v_r^{-1}[q_{t,k}(k) - q_r(k)]^{-1} + w_k{}'P_z^{-1}w_k\} \\
&\quad + 2[q_{t,k}(k) - q_r(k)]\, \mathrm{L}n[\lambda/(1 - \lambda)] \\
&\quad - \mathrm{L}n\{1 + w_k{}'P_z^{-1}w_k\, v_r[q_{t,k}(k) - q_r(k)]\}\ ,\ k = 1, 2, \ldots , N
\end{aligned}
\tag{8-35}
$$

Comparing $\mathrm{L}n D^1(z;k)$ in Eq. (8-35) with $\mathrm{L}n D(z;k)$ in Eq. (5-9), we observe that

$$
\mathrm{L}n D^1(z;k) = \mathrm{L}n D(z;k) - \mathrm{L}n\{1 + w_k{}'P_z^{-1}w_k\, v_r[q_{t,k}(k) - q_r(k)]\}
\tag{8-36}
$$

The second term in Eq. (8-36) has a tremendous stabilizing effect on the first term, so that the SMLR[1] detector does not have the false alarm problems that are associated with the SMLR detector. Recall, however, from Example 6-7, that even the SMLR detector can be 'stabilized' by the simple inclusion of backscatter.

As pointed out by Mendel (1983), it is numerically possible for the argument of the $\mathrm{L}n\{\ \}$ term in Eq. (8-35) to become negative. Remember that the derivation of the SMLR[1] detector assumes that we have access to the true wavelet and statistical parameters. If we had these true parameters, the argument of the $\mathrm{L}n\{\ \}$ term would not become negative. When the SMLR[1] detector is used in the block component method, we estimate these parameters and the just-mentioned numerical problem may occur. If it does occur, ignore the time point at which it happens, and proceed to the next time point (i.e., the next value of k) in the SMLR[1] likelihood-ratio decision function $\mathrm{L}n D^1(z;k)$.

8.8 Quadratic Convergence of the Newton-Raphson Algorithm

Here we prove that if $f(x)$ is itself a quadratic function, we obtain the optimal solution x^* in a single step of the Newton-Raphson algorithm. If $f(x)$ is quadratic, then

$$
f(x) = -x'Ax/2
\tag{8-37}
$$

The Newton-Raphson algorithm for updating x, so as to maximize $f(x)$, is (e.g., Fox, 1971)

$$
x_{i+1} = x_i - H_{x,i}^{-1} g_{x,i} \qquad i = 0, 1, \ldots
\tag{8-38}
$$

where $g_{x,i}$, the gradient of $f(x)$ evaluated at x_i, equals $-Ax_i$, and $H_{x,i}$, the Hessian matrix of $f(x)$ evaluated at x_i, equals $-A$. Substituting these quantities into Eq. (8-38), we find that $x_1 = x_0 - A^{-1}(Ax_0) = 0$. We know ahead of time that the maximum value of $f(x) = -x'Ax/2$ occurs at $x = 0$; thus, because $x_1 = 0$, regardless of the value of x_0, we achieve the maximum of $f(x)$ in exactly one step of the Newton-Raphson algorithm.

8.9 Wavelet Identifiability

Here we show that when the event sequence is known to be a Bernoulli sequence for which $\lambda \neq 0$ and $\lambda \neq 1$, then $\Phi_{qq}(\omega) \neq 0$ for all ω and we are therefore able to identify wavelet w(k) to within a scale factor of its true value. This result is taken from Kormylo and Mendel (1983) and Mendel (1983).

The question to be answered is 'Will $\Phi_{qq}(\omega) \neq 0$ for all ω when q(k) is a Bernoulli sequence?' As pointed out in Chapter 5, when q(k) is known to be a Bernoulli sequence, it is ergodic in correlation; hence, $\Phi_{qq}(\omega)$ can be calculated using Eq. (5-20). We must compute the autocorrelation, $r_{qq}(l)$, of q(k) and then evaluate the discrete Fourier transform of $r_{qq}(l)$.

Because $q^2(i) = q(i)$ and q(i) and q(i + l) are statistically independent for all $i \neq i + l$, we see that

$$E\{q(i)q(i+l)\} = \begin{cases} E\{q(i)\} = \lambda & \text{when } l = 0 \\ E\{q(i)\}E\{q(i+l)\} = \lambda^2 & \text{when } l \neq 0 \end{cases} \tag{8-39}$$

This result can be expressed more succinctly as

$$r_{qq}(l) = (\lambda - \lambda^2)\delta(l) + \lambda^2 \tag{8-40}$$

Finally, we compute the discrete Fourier transform of $r_{qq}(l)$, as

$$\Phi_{qq}(\omega) = \sum_{l=-\infty}^{\infty} [(\lambda - \lambda^2)\delta(l) + \lambda^2] e^{-j\omega l} \tag{8-41}$$

or

$$\Phi_{qq}(\omega) = (\lambda - \lambda^2) + \lambda^2 \delta(\omega) \tag{8-42}$$

As long as $\lambda \neq 0$ or $\lambda \neq 1$, $\Phi_{qq}(\omega) \neq 0$ for *all* ω, and we therefore satisfy Eq. (5-18). Consequently, in the seismic case, where q(k) is Bernoulli, we are indeed able to identify wavelet w(k) to within a scale factor of its true value.

8.10 Convergence of an Adaptive SMLR Detector

Alternating iterations of the SMLR detector and the update for event location parameter λ, given in Eq. (4-22), leads to an adaptive SMLR detector. It is adaptive in the sense that parameter λ, which is used by the SMLR detector, changes from iteration to iteration. Here we shall show that this adaptive SMLR detector converges in a finite number of steps. This result is taken from Kormylo (1979) and Mendel (1983).

For a constant value of λ, each iteration of the SMLR detector increases function $\mathcal{M}$. Similarly, for constant q, update (4-22) also increases function $\mathcal{M}$. Consequently, alternating update (4-22) and the SMLR detector produces a monotonic increase in likelihood. Because there are only a finite number of sequences q (namely, 2^N) and corresponding values for

$$m = \sum_{k=1}^{N} q_{SMLR}(k) \,,$$

this two-stage algorithm must converge in a finite number of steps.

9
Computational Considerations

9.1 Introduction

The entire theory of Maximum-Likelihood Deconvolution has been developed in the context of the familiar convolutional model. While this has expedited the development of all its results, it has unfortunately led us to algorithms that are very impractical to use on today's digital computers. This was already mentioned, in part, in Chapter 7.

Consider the Block Component Method depicted in Figure 4-20, which we repeat here for the convenience of the reader as Figure 9-1. Table 9-1 summarizes the gross computational requirements for each component of this Block Component Method. This is done for both $\mathcal{L}$ and $\mathcal{M}$.

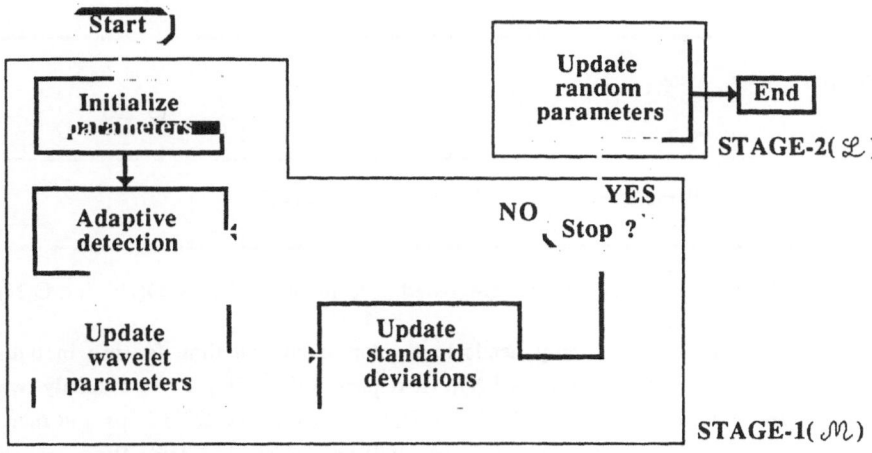

Figure 9-1. Two-stage likelihood optimization algorithm with an adaptive detector.

Table 9-1. Computational Requirements for Figure 9-1 Block Component Method: Batch Calculations.

COMPONENT	EQUATIONS	
Initialize Parameters		
MVD (needed for Threshold Detector)	(7-35)&(7-38)[b]	
Threshold Detector	(7-62) to (7-64)	
Adaptive Detection		
SMLR Detector	(7-93)&(7-95)[c]	
λ-Update	(7-184)	
Update Wavelet Parameters	$\mathcal{M}$	$\mathcal{L}$
Gradients	12step algorithm[d]	9 step algorithm[e]
Pseudo-Hessians	(7-159) to (7-161)	(7-166) to (7-169)
Marquardt-Levenberg[f]	(4-8)&(4-9)	(4-8)&(4-9)
Update Standard Deviations	$\mathcal{M}$	$\mathcal{L}$
Gradients	5 step algorithm	4 step algorithm[e]
Pseudo-Hessians	(7-170)&(7-171)	(7-174)&(7-175)
Marquardt-Levenberg	(4-8)&(4-9)	(4-8)&(4-9)
Stop?[g]	$\mathcal{M}$	$\mathcal{L}$
	(7-139)	(5-22)
Update Random Parameters	(7-35)	

a. Only $O(N^3)$ calculations are listed. A blank indicates $O(N^2)$ or $O(N)$ calculations.
b. $(WP_uW' + v_nI)^{-1}$ only needs to be computed one time for use in both $u^{MV}(k \mid N)$ and $Var[u_{ERR}(k \mid N)]$. It requires $O(N^3)$ flops. Additionally, we are assuming that $w_k'(WP_uW' + v_nI)^{-1} w_k$ requires $O(2N^2)$ flops, but there are N of these calculations, so there will be a total of $O(2N^3)$ flops needed to compute $Var[u_{ERR}(k \mid N)]$. If the structure of W is used in coding $(WP_uW' + v_nI)^{-1}$, much fewer flops are required.
c. See also (7-98) and (7-99). MVD is needed by the SMLR detector.
d. Steps 10 and 11 are the computationally intensive ones. They require solutions of $(2n+1)$ N×N linear equations.

Table 9-1 (continued)

FLOPS[a]	
$3N^3$	
--	
$3N^3$	
--	
$\mathcal{M}$	$\mathcal{L}$
$(2n + 1)N^3$	N^3
$n(n - 1)N^3$	--
--	--
$\mathcal{M}$	$\mathcal{L}$
$3N^3$	N^3
--	--
--	--
$\mathcal{M}$	$\mathcal{L}$[h]
N^3 per calculation	
N^3	

e. Requires (at least) one calculation of $u^{MV}(k \mid N)$ that is used in *both* gradient and pseudo-Hessian calculations.

f. $(H_{a,i} + D_{a,i})^{-1}$ and $(H_{b,i} + D_{b,i})^{-1}$ each require $O(n^3)$ flops and $n \ll N$.

g. If this BCM is programmed for an interactive environment, in which the user chooses how many iterations are spent in each block, then $\mathcal{M}$ and $\mathcal{L}$ are usually computed for each iteration in a block and are displayed to the user, so that the user can decide whether or not it is worthwhile to iterate further in the block.

h. Requires a calculation of y from (7-138) each time $\mathcal{M}$ is computed, or $u^{MV}(k \mid N)$ each time $\mathcal{L}$ is computed.

Assuming α entries into the Adaptive Detection block, β iterations in the Update Wavelet Parameters block, γ iterations in the Update Standard Deviations block, and one calculation of $\mathcal{L}$ or $\mathcal{M}$ per iteration within each block (which adds another N^3 flops to each stay in a block), the total gross numbers of flops when we use $\mathcal{M}$ or $\mathcal{L}$, denoted $\mathrm{Flops_B}(\mathcal{M})$ and $\mathrm{Flops_B}(\mathcal{L})$, where the subscript 'B' denotes 'batch' calculations, are

$$\begin{aligned} \mathrm{Flops_B}(\mathcal{M}) &= (3N^3 + N^3) + \alpha(3N^3 + N^3) + \beta[(n^2 + n + 1)N^3 + N^3] \\ &\quad + \gamma(3N^3 + N^3) + (N^3 + N^3) \\ &= 4N^3\alpha + [(n^2 + n)N^3 + 2N^3]\beta + 4N^3\gamma + 6N^3 \end{aligned} \tag{9-1}$$

and

$$\begin{aligned} \mathrm{Flops_B}(\mathcal{L}) &= (3N^3 + N^3) + \alpha(3N^3 + N^3) + \beta(N^3 + N^3) \\ &\quad + \gamma(N^3 + N^3) + (N^3 + N^3) \\ &= 4N^3\alpha + 2N^3\beta + 2N^3\gamma + 6N^3 \end{aligned} \tag{9-2}$$

Observe that the major difference between $\mathrm{Flops_B}(\mathcal{M})$ and $\mathrm{Flops_B}(\mathcal{L})$ is in the factor $(n^2 + n)N^3$ in $\mathrm{Flops_B}(\mathcal{M})$, which is associated with the calculation of the $2n$ gradients and pseudo-Hessians of $\mathcal{M}$ with respect to the wavelet's parameters.

As an example to illustrate these results, suppose $N = 1,000$, $n = 6$ (i.e., we have a 6th-order ARMA wavelet), and, $\alpha = \beta = \gamma = 1$. Then $\mathrm{Flops_B}(M) = 58 \times 10^9$ and $\mathrm{Flops_B}(\mathcal{L}) = 14 \times 10^9$. If 1 flop requires 10^{-6} sec. of computation time, then $\mathrm{Flops_B}(\mathcal{M})$ requires 58×10^3 sec. $= 16.11$ hr., whereas $\mathrm{Flops_B}(\mathcal{L})$ requires 14×10^3 sec. $= 3.89$ hr. Clearly, neither of these huge computation times is acceptable. Remember, also, that the Block Component Method requires many iterations within the Adaptive Detection, Update Wavelet Parameters and Update Standard Deviations blocks; hence, the actual computational picture is significantly worse than the dismal picture already portrayed. *An alternative to batch processing must be used!*

What are the alternatives that are available to us today? *Parallel processing* is one alternative. This has been briefly described in the Chapter 7 sections entitled "Calculating Gradients: Gradients of $\mathcal{M}$ with respect to **a** and **b**," and "Calculating Gradients: Gradients of $\mathcal{L}$ with respect to **a** and **b**." See, for example, Figures 7-8 and 7-9. In these sections it is conjectured that a 1,000-fold savings in computation time can be achieved using parallel processing. For the preceding example, this would mean that $\mathrm{Flops_B}(\mathcal{M})$ would require 58sec., whereas $\mathrm{Flops_B}(\mathcal{L})$ would require 14 sec. These computation times are still unacceptable, especially in a highly interactive environment. Of course, if an even greater savings in computation time can be achieved using parallel processing, then batch processing of the data would be practical.

A second alternative is to use a shorter time window of data. This can be accomplished either by sampling at a slower rate (e.g., 2 msec vs 1 msec) or by just using fewer data points at a fixed sampling rate. For example, if $N_2 = 500$

instead of $N_1 = 1000$, then $(N_2/N_1)^3 = 0.125$. Calculations can then be done in roughly 1/8th the time they previously could have been done in. For batch processing, this means that $\text{Flops}_B(\mathcal{M})$ now requires 2.104 hr., whereas $\text{Flops}_B(\mathcal{L})$ now requires .486 hr. These times are still unacceptable. Parallel processing might further reduce these times to 7.25 sec. and 1.75 sec., respectively. Further reducing the number of data points will, of course reduce computation times, but at the expense of resolution and accuracy. Taken to the absurd limit, if we process no data then we require zero computation time, and, of course, we learn nothing!

A third alternative to batch processing is *recursive processing*. The main purpose of this chapter is to present, without derivations, a collection of recursive signal processing algorithms for MVD, etc. The savings in computation time, when these algorithms are run on serial processing digital computers, is astounding. If these algorithms are run on a parallel computer, they would then enjoy whatever speedup factor the batch algorithms would enjoy. Hence, as mentioned in Chapter 7, there may be no reason to ever want to compute with the batch formulas.

9.2 Recursive Processing

Throughout this book we have assumed an ARMA model for the wavelet; however, except for calculating gradients and pseudo-Hessians of $\mathcal{L}$ and $\mathcal{M}$ (in Chapter 7) we have never fully exploited the recursive nature of that model. The recursive signal processing algorithms that we shall present in this section exploit this feature of an ARMA model to its fullest.

9.2.1 A Recursive Wavelet Model

Recall, from Chapter 2, that the z-transform transfer function of an ARMA wavelet is

$$W(z) = (b_1 z^{n-1} + b_2 z^{n-2} + \ldots + b_{n-1} z + b_n)/(z^n + a_1 z^{n-1} + \ldots + a_{n-1} z + a_n) \quad (9\text{-}3)$$

When z denotes the unit advance operator [i.e., $zf(t) = f(t + 1)$], then this transfer function implies the following *ARMA finite-difference equation*:

$$y(k + n) \ + a_1 y(k + n - 1) + \ldots + a_{n-1} y(k + 1) + a_n y(k) = b_1 u(k + n - 1)$$
$$+ b_2 u(k + n - 2) + \ldots + b_{n-1} u(k + 1) + b_n u(k) \quad (9\text{-}4)$$

To obtain this result we have used (9-3) and the fact, from Figure 2-13, that $Y(z) = W(z)U(z)$. Equation (9-4) can also be written as

$$y(k + n) = -a_1 y(k + n - 1) - ... - a_{n-1} y(k + 1) - a_n y(k) + b_1 u(k + n - 1)$$
$$+ b_2 u(k + n - 2) + ... + b_{n-1} u(k + 1) + b_n u(k) \qquad (9\text{-}5)$$

which demonstrates that output y at time $k + n$ ($k = 0, 1, ...$) can be calculated *recursively* from the earlier values of y, namely, $y(k + n - 1), ... , y(k)$, and, the input values $u(k + n - 1), ... , u(k)$.

Equation (9-5) is an nth-order finite-difference equation. It can also be expressed as a collection of n first-order finite-difference equations. These n equations can then be collected together , using vector and matrix notations, into a more compact form known as a *state equation*.

Example 9-1. Consider the third-order ARMA model for which

$$W(z) = (b_1 z^2 + b_2 z + b_3)/(z^3 + a_1 z^2 + a_2 z + a_3) = Y(z)/U(z) \qquad (9\text{-}6)$$

Introduce intermediate variable $y_1(k)$ whose z transform is $Y_1(z)$:

$$W(z) = (b_1 z^2 + b_2 z + b_3)Y_1(z)/(z_3 + a_1 z^2 + a_2 z + a_3)Y_1(z) = Y(z)/U(z) \qquad (9\text{-}7)$$

Equate numerator and denominator terms to see that

$$y(k) = b_1 y_1(k + 2) + b_2 y_1(k + 1) + b_3 y_1(k) \qquad (9\text{-}8)$$

and

$$y_1(k + 3) + a_1 y_1(k + 2) + a_2 y_1(k + 1) + a_3 y_1(k) = u(k) \qquad (9\text{-}9)$$

Observe, from these last two equations, that output $y(k)$ is now related to the 'output' $y_1(k)$ of an AR model.

Equation (9-9) is a 3rd-order AR model. We shall now reexpress it as a collection of three first-order difference equations. There are many ways to do this. Here is one of the simplest ones. Let

$$x_1(k) \triangleq y_1(k), \quad x_2(k) \triangleq y_1(k + 1) \text{ and } x_3(k) \triangleq y_1(k + 2) \qquad (9\text{-}10)$$

Then

$$x_1(k + 1) = y_1(k + 1) = x_2(k)$$

$$x_2(k + 1) = y_1(k + 2) = x_3(k)$$

$$x_3(k + 1) = y_1(k + 3) = - a_1 y_1(k + 2) - a_2 y_1(k + 1) - a_3 y_1(k) + u(k)$$
$$= - a_1 x_3(k) - a_2 x_2(k) - a_3 x_1(k) + u(k)$$

These three equations can also be collected into the following vector-matrix representation:

$$
\begin{pmatrix} x_1(k+1) \\ x_2(k+1) \\ x_3(k+1) \end{pmatrix} = \begin{pmatrix} 0 & 1 & 0 \\ 0 & 0 & 1 \\ -a_3 & -a_2 & -a_1 \end{pmatrix} \begin{pmatrix} x_1(k) \\ x_2(k) \\ x_3(k) \end{pmatrix}
$$

$$
+ \begin{pmatrix} 0 \\ 0 \\ 1 \end{pmatrix} u(k) \tag{9-11}
$$

Additionally, because $y_1(k) = x_1(k)$,

$$
y_1(k) = (\,1 \quad 0 \quad 0\,) \begin{pmatrix} x_1(k) \\ x_2(k) \\ x_3(k) \end{pmatrix} \tag{9-12}
$$

Equations (9-11) and (9-12) constitute the state-variable representation for the AR model in Eq. (9-9).

Finally, our desired ARMA output $y(k)$ can be expressed in terms of the three state variables, $x_1(k)$, $x_2(k)$, and $x_3(k)$, using Eq. (9-8), as

$$
y(k) = (\,b_3 \quad b_2 \quad b_1\,) \, x(k) \tag{9-13}
$$

where $x(k) = \text{col}[x_1(k), x_2(k), x_3(k)]$ is our state vector. Equations (9-11) and (9-13) constitute the state-variable representation for the ARMA model in Eq. (9-6).

Observe that the ARMA model reduces to an AR model by setting $b_1 = b_2 = 0$ or to a MA model by setting $a_1 = a_2 = a_3 = 0$.□

The results in this example are easily generalized. A state-variable representation for the ARMA model in Eq. (9-3) is

$$
\begin{pmatrix} x_1(k+1) \\ x_2(k+1) \\ \cdot \\ \cdot \\ \cdot \\ x_n(k+1) \end{pmatrix} = \begin{pmatrix} 0 & 1 & 0 & \cdots & 0 \\ 0 & 0 & 1 & \cdots & 0 \\ \cdot & \cdot & \cdot & \cdot & \cdot \\ \cdot & \cdot & \cdot & \cdot & \cdot \\ \cdot & \cdot & \cdot & \cdot & \cdot \\ -a_n & -a_{n-1} & -a_{n-2} & \cdots & -a_1 \end{pmatrix} \begin{pmatrix} x_1(k) \\ x_2(k) \\ \cdot \\ \cdot \\ \cdot \\ x_n(k) \end{pmatrix}
$$

$$
+ \begin{pmatrix} 0 \\ 0 \\ \cdot \\ \cdot \\ \cdot \\ 1 \end{pmatrix} u(k) \tag{9-14}
$$

and

$$
y(k) = (b_n, b_{n-1}, \ldots, b_1)x(k) \tag{9-15}
$$

Remember that Eq. (9-14) is nothing more than a collection of n first-order finite-difference equations that have been collected together and packaged into a vector-matrix format. This equation is called a *state equation*, whereas Eq. (9-15) is called an *output equation*, because y(k) is indeed the output of our wavelet model.

Equations (9-14) and (9-15) can be written in the following more compact notation:

$$
x(k+1) = \Phi x(k) + \gamma u(k) \tag{9-16}
$$

and

$$
y(k) = h'x(k) \tag{9-17}
$$

In these equations: x(k) is always treated as an $n \times 1$ state vector; Φ is an $n \times n$ state transition matrix; γ is an $n \times 1$ input distribution vector (i.e., it 'distributes' the input into those n difference equations in which the input should appear); and, h is an $n \times 1$ observation vector. Of course, examples of Φ, γ and h can be deduced for the ARMA model by comparing Eqs. (9-16) and (9-17) with Eqs. (9-14) and (9-15), respectively. Some examples of non-ARMA wavelet models can be found in Mendel and Kormylo (1978).

Equations (9-16) and (9-17) constitute a single-channel state-variable representation for the convolutional model $y(k) = u(k)*w(k)$. The state equation model provides an *internal representation* for the wavelet model. It is important to understand that it provides exactly the same relationship between u(k) and y(k) as does the convolutional model; it just expresses this relationship in a different mathematical way (see Figure 9-2). Although we have derived it for an ARMA wavelet model, it is by no means limited to such a model, nor is it limited to single-channel models, or to time-invariant wavelets. For more details about state space models, see Mendel, et al. (1981), Mendel (1983), or a standard textbook on linear systems, such as Gabel and Roberts (1980). Other state-variable models that have been advocated to achieve "fast" minimum-variance deconvolution are described in Demoment, et al. (1984), Demoment and Reynaud (1985), Deng (1985), and Goussard and Demoment (1987).

Equations (9-16) and (9-17) are the bases for the recursive signal processing algorithms that we shall present next.

9.2.2   Recursive MVD Algorithm

From our earlier chapters, it is clear that the MVD algorithm is at the very heart of the entire MLD procedure. A recursive MVD algorithm is derived most directly from mean-squared estimation theory applied to our state-variable model. Entire textbooks have been written on the subject of mean-squared estimation theory [e.g., Meditch (1969), Anderson and Moore (1979), and Mendel (1987)]. Much has also been written on deriving MVD algorithms from this theory [e.g., Mendel (1983, 1987)]. Here we just state what this author believes to be the most useful version of a recursive MVD algorithm. It consists of four components.

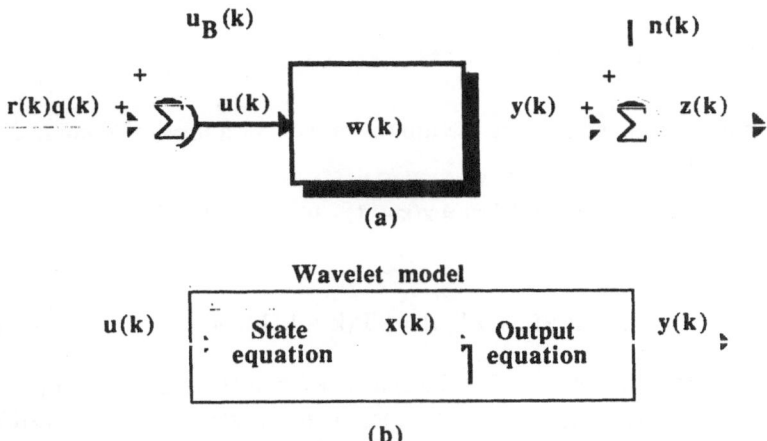

Figure 9-2. (a) convolutional model; (b) interpretation of wavelet model as an interconnection of a state equation and output equation.

9.2.2.1 Input Estimator

Let $r(j \mid N)$ denote an $n \times 1$ backward-running (i.e., $j = N, N - 1, \ldots , 0$) state vector, and $S(j \mid N)$ denote $r(j \mid N)$'s associated $n \times n$ covariance matrix. Then

$$u^{MV}(k \mid N) = v_u \gamma \,' r(k + 1 \mid N) \tag{9-18}$$

$$\mathrm{Var}[u_{ERR}(k \mid N)] = v_u - v_u \gamma \,' S(k + 1 \mid N) \gamma \, v_u \tag{9-19}$$

where $k = N, N - 1, \ldots , 1$.

9.2.2.2 Backward-Running Filter

The backward-running state and covariance equations for $r(j \mid N)$ and $S(j \mid N)$ are:

$$r(j \mid N) = [I - K(j)h']' \Phi' r(j + 1 \mid N) + h\tilde{z}(j \mid j - 1) / \mathrm{Var}[\tilde{z}(j \mid j - 1)] \tag{9-20}$$

and

$$S(j \mid N) = [I - K(j)h']' \Phi' S(j + 1 \mid N) \Phi [I - K(j)h'] + hh' / \mathrm{Var}[\tilde{z}(j \mid j - 1)] \tag{9-21}$$

where $j = N, N - 1, \ldots , 1$, $r(N + 1 \mid N) = 0$ and $S(N + 1 \mid N) = 0$. In these equations $\tilde{z}(j \mid j - 1)$ is a scalar process known as the *innovations* [given below in Eq. (9-22)], and $K(k)$ is an $n \times 1$ gain matrix, known as the *Kalman gain matrix* [given below in Eq. (9-27)].

9.2.2.3 Innovations Process

The innovations and its variance are the outputs of a forward-running *Kalman predictor*, i.e.,

$$\tilde{z}(k + 1 \mid k) = z(k + 1) - h'\hat{x}(k + 1 \mid k) \tag{9-22}$$

and

$$\mathrm{Var}[\tilde{z}(k + 1 \mid k)] = h'P(k + 1 \mid k)h + v_n \tag{9-23}$$

where $k = 0, 1, \ldots , N - 1$; $\hat{x}(k + 1 \mid k)$ denotes an $n \times 1$ mean-squared prediction of state vector $x(k + 1)$, based on the measurements $z(1), z(2), \ldots$, and $z(k)$; and, $P(k + 1 \mid k)$ is the $n \times n$ covariance matrix that describes the estimation error between $\hat{x}(k + 1 \mid k)$ and $x(k + 1)$. Quantities $\hat{x}(k + 1 \mid k)$ and $P(k + 1 \mid k)$ are generated from the Kalman predictor.

9.2.2.4 Kalman Predictor

Let $\hat{x}(k \mid k)$ and $P(k \mid k)$ denote an $n \times 1$ mean-squared *filtered* estimate of state vector $x(k)$, one that is based on the measurements $z(1)$, $z(2)$, ... , and $z(k)$, and, its associated $n \times n$ covariance matrix. Matrix $P(k \mid k)$ describes the estimation error between $\hat{x}(k \mid k)$ and $x(k)$. Then

$$\hat{x}(k + 1 \mid k) = \Phi\hat{x}(k \mid k) \qquad (9\text{-}24)$$

and

$$P(k + 1 \mid k) = \Phi P(k \mid k)\Phi' + v_u \gamma\gamma' \qquad (9\text{-}25)$$

where

$$\hat{x}(k + 1 \mid k + 1) = [I - K(k + 1)h']\hat{x}(k + 1 \mid k) + K(k + 1)z(k + 1) \qquad (9\text{-}26)$$

$$K(k + 1) = P(k + 1 \mid k)h[h'P(k + 1 \mid k)h + v_n]^{-1} \qquad (9\text{-}27)$$

$$P(k + 1 \mid k + 1) = [I - K(k + 1)h']P(k + 1 \mid k) \qquad (9\text{-}28)$$

and $k = 0, 1, ... , N - 1$. The Kalman predictor can be initialized by $\hat{x}(0 \mid 0) = 0$ and $P(0 \mid 0) = 0$.

Figure 9-3 depicts the interconnection of these four computational subsystems. Observe that they are interconnected in a reverse order from the order in which we presented the algorithm. If you begin at the output of the algorithm, then the ordering is the one we have stated. Observe, also, that the *noncausal* nature of the MVD algorithm is preserved in this recursive implementation.

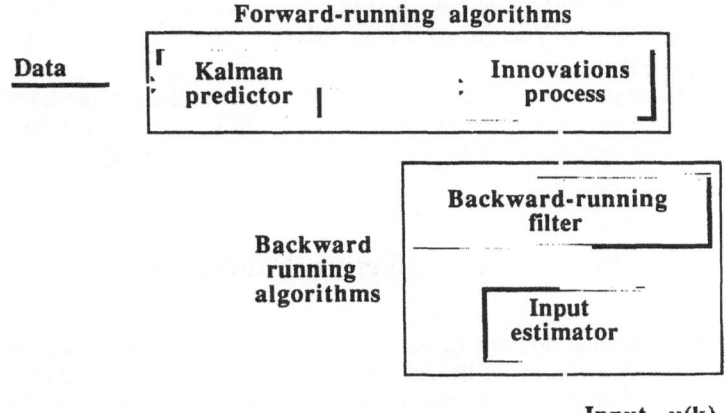

Figure 9-3. Interconnection of the four computational subsystems that comprise the recursive MVD algorithm.

There are no multiplications of N×N matrices in the recursive MVD filter. The most intensive calculations appear in Eqs. (9-21) and (9-25), and are the n×n matrix products $\Phi'S(j + 1 \mid N)\Phi$ and $\Phi P(k \mid k)\Phi'$, each of which requires on the order of $2n^3$ flops per iteration (and, there are N iterations). The order of the ARMA wavelet, n, is vastly smaller than data length N. A more detailed computation count is given at the end of this chapter.

Many other recursive MVD algorithms have appeared in the literature. Some of these algorithms are based on a different wavelet state-variable model [Demoment and Reynaud (1985), Demoment, et. al. (1984),and Deng (1985)]; others are based on techniques from system identification [Kollias and Halkias (1985)]; others do not have to use all the N measurements (e.g., they only have to use a fixed window of measurements, where the window is much smaller than N) [Kollias, et. al. (1985), Kollias and Halkias (1985), and Lainiotis, et. al. (1988)]; and, others permit the user to use radically different recursive formulas, hopefully to achieve reduced computations [Lainiotis, et. al. (1988)].

Recently, even neural network implementations of MVD have appeared (Zhao and Mendel, 1988a,b); however, these are nonrecursive.

9.2.3   Detection

Both the threshold and SMLR detectors use information from MVD filtering. Regarding the former, refer to Eqs. (7-62) - (7-64), to see that both $u^{MV}(k \mid N)$ and $Var[u_{ERR}(k \mid N)]$ are used to implement the threshold detector. Regarding the latter, refer to Eqs. (7-93), (7-95), (7-98) and (7-99), to see that both $u^{MV}(k \mid N)$ and $Var[u_{ERR}(k \mid N)]$ are also used to implement the SMLR detector. The recursive MVD algorithm we have just presented can now be used to compute both $u^{MV}(k \mid N)$ and $Var[u_{ERR}(k \mid N)]$, thereby reducing the computational burden of these detectors.

All of our detectors operate in an off-line, non real-time modality. For detectors that operate in a more real-time modality, see Kormylo (1979), Goussard and Demoment (1987b) and Mahalanabis, et. al. (1982).

9.2.4   Likelihood and Objective Functions

In order to implement any one of our Block Component Methods, loglikelihood or objective functions must be calculated. This is done to be sure that the loglikelihood or objective function is being maximized as we move from one block to the next. These functions can be calculated using recursive processing algorithms, as we demonstrate in this subsection.

Consider $\mathcal{L} = \ln L$ first, where, from Eq. (3-2)

$$\mathscr{L}\{a, b, s, q, r, u_B \mid z\} = \text{ln}p(z \mid q, r, u_B, a, b, s) + \text{ln}Pr(q \mid s)$$
$$+ \text{ln}p(r \mid s) + \text{ln}p(u_B \mid s) \qquad (9\text{-}29)$$

The most complicated part of this loglikelihood function is $\text{ln}p(z \mid q, r, u_B, a, b, s)$; equations for $\text{ln}Pr(q \mid s)$, $\text{ln}p(r \mid s)$, and $\text{ln}p(u_B \mid s)$ are easily determined from Eqs. (3-5), (3-7) and (3-8), respectively.

Referring to our state-variable model in Eqs. (9-16) and (9-17), augmented by the fact that we only have access to a noisy measurement, $z(k)$, where

$$z(k) = y(k) + n(k) = h'x(k) + n(k), \quad k = 1, 2, \dots, N \qquad (9\text{-}30)$$

we observe that, when q, r, and u_B are given, there are only two possible sources of uncertainty (i.e., randomness) left in our model, namely $n(k)$ and the initial state vector $x(0)$. For our particular convolutional model, $x(0)$ *must equal zero*, or else the solution of the state equation [and, therefore, $y(k)$] will contain a transient term that depends upon $x(0)$ as well as the convolutional term (Mendel, 1983). Consequently, the only remaining source of uncertainty is the *white* Gaussian measurement noise process $n(k)$. State vector $x(k)$ is deterministic when q, r, and u_B are given. Because of these facts,

$$p(z \mid q, r, u_B, a, b, s) = \prod_{i=1}^{N} p(z(i) \mid q, r, u_B, a, b, s) \qquad (9\text{-}31)$$

so that

$$\text{ln}p(z \mid q, r, u_B, a, b, s) = \sum_{i=1}^{N} \text{ln}p(z(i) \mid q, r, u_B, a, b, s) \qquad (9\text{-}32)$$

From the Gaussian nature of $n(k)$ and the linear measurement model in Eq. (9-30), we know that

$$p(z(i) \mid q, r, u_B, a, b, s)$$
$$= [1/(\sqrt{2\pi}\,v_n)]\,\text{exp}\{-\tfrac{1}{2}[z(i) - h'x(i)]'[z(i) - h'x(i)]\}\ ; \qquad (9\text{-}33)$$

consequently,

$$\text{ln}p(z \mid q, r, u_B, a, b, s) = -\tfrac{1}{2}\sum_{i=1}^{N} [z(i) - h'x(i)]'[z(i) - h'x(i)]/v_n$$
$$- \tfrac{N}{2}\,\text{ln}v_n - \tfrac{N}{2}\,\text{ln}2\pi \qquad (9\text{-}34)$$

Combining Eq. (9-34) with the logarithmic versions of Eqs. (3-5), (3-7) and (3-8), we obtain the final result, that

$$\mathcal{L}\{a, b, s, q, r, u_B \mid z\} = -\frac{N}{2}\ln 2\pi - \frac{1}{2}\sum_{i=1}^{N}[z(i) - h'x(i)]'[z(i) - h'x(i)]/v_n$$

$$-\frac{N}{2}\ln v_n v_r v_B - r'r/2v_r - u_B'u_B/2v_B$$

$$+ m(q)\ln(\lambda) + [N - m(q)]\ln(1 - \lambda) \qquad (9\text{-}35)$$

Comparing Eqs. (9-35) and (5-22), we see that they differ only in the ways in which the effects of the wavelet model are calculated. In Eq. (5-22) the calculations are based on the convolutional model, whereas in Eq. (9-35) the calculations are based on the state-variable model. Observe that to calculate $h'x(i)$, state vector $x(i)$ must be computed by solving the state equation (9-16). Of course, this is done in a recursive manner; hence, the state equation is coupled into loglikelihood function $\mathcal{L}\{a, b, s, q, r, u_B \mid z\}$. In effect, it acts as a constraint on $\mathcal{L}$, when we go to maximize $\mathcal{L}$.

Next, we consider $\mathcal{M} = \ln \mathcal{M}$. Recall that M is not a likelihood function; it is a function that is closely related to L, but is obtained as a result of the Separation Principle. Hence, we cannot obtain a different formula for M in a direct manner, as we just did for L, by applying probability theory to our state-variable model. Recall, however, that N is a likelihood function (see the section in Chapter 5, entitled "A Modified Likelihood Function"). As pointed out at the end of Chapter 5, there is a very strong relationship between N and M. Recall that [see Eqs. (5-29) and (5-30)]

$$M\{a, b, s, q \mid z\} = (2\pi)^{-N}\mid \Omega \mid^{1/2}(v_n v_r v_B)^{-N/2} N\{a, b, s, q \mid z\} ; \qquad (9\text{-}36)$$

hence,

$$\mathcal{M}\{a, b, s, q \mid z\}$$
$$= -N\ln 2\pi - \frac{N}{2}\ln v_n v_r v_B + \frac{1}{2}\ln \mid \Omega \mid + \mathcal{N}\{a, b, s, q, z\} \qquad (9\text{-}37)$$

In order to obtain a recursive procedure for calculating M we must first obtain such a procedure for N. The details for doing this require the following facts about the innovations process: (1) $\tilde{z}(k \mid k - 1)$ contains the same information as does the measurement $z(k)$, (2) $\tilde{z}(k \mid k - 1)$ is a white noise process, and (3) $\tilde{z}(k \mid k - 1)$ is Gaussian. The proofs of these facts can be found in books on estimation theory [e.g., Anderson and Moore (1979) and Mendel (1987)]. The first fact is sometimes stated as "$\tilde{z}(k \mid k - 1)$ and $z(k)$ are causally invertible." This means that it is always possible to determine $\tilde{z}(k \mid k - 1)$ from $z(k)$ and vice versa using causal transformations (i.e., forward-running filters). The second fact means that $\tilde{z}(i \mid i - 1)$ and $\tilde{z}(j \mid j - 1)$ are uncorrelated at all values of $i \neq j$; hence, the cross-covariance matrix between $\tilde{z}(i \mid i - 1)$ and $\tilde{z}(j \mid j - 1)$ is a diagonal matrix. The elements along the diagonal of this matrix are $Var[\tilde{z}(i \mid i - 1)]$. The third fact, which is due to the Gaussian nature of $r(k)$, $u_B(k)$ and $n(k)$, lets us write down a simple formula for a conditional density function of $\tilde{z}(k \mid k - 1)$.

Before we state a new formula for loglikelihood function $\mathcal{N}\{a, b, s, q \mid z\}$, the reader should return to Eq. (8-26) in the section in Chapter 8, entitled "Modified Likelihood Function," to observe that the most difficult part of the calculation of $\mathcal{N}$ is $\ln p(z \mid q, a, b, s)$. Note that the conditioning with respect to the random event vector q does *not* remove all sources of uncertainty from the convolutional model. The uncertainties with respect to the amplitude vector r and backscatter vector u_B are still present; hence, in this case state vector $x(k)$ is random due to $r(k)$ and $u_B(k)$. Unlike the previous case, where

$$p(z \mid q, r, u_B, a, b, s) = \prod_{i=1}^{N} p(z(i) \mid q, r, u_B, a, b, s) ,$$

in the present case

$$p(z \mid q, a, b, s) \neq \prod_{i=1}^{N} p(z(i) \mid q, a, b, s) \tag{9-38}$$

because the measurements are all correlated due to their dependence on the now-random state vector $x(k)$.

The causal invertibility between $\tilde{z}(k \mid k - 1)$ and $z(k)$ permits us to replace $z(k)$ by $\tilde{z}(k \mid k - 1)$ in the derivation of $\mathcal{N}\{a, b, s, q \mid z\}$. The big advantages for doing this, are

$$p(\tilde{z} \mid q, a, b, s) = \prod_{i=1}^{N} p(\tilde{z}(i \mid i - 1) \mid q, a, b, s), \tag{9-39}$$

and, it is relatively straightforward to compute $p(\tilde{z}(i \mid i - 1) \mid q, a, b, s)$. The final expression for $\mathcal{N}\{a, b, s, q \mid z\}$ is

$$\mathcal{N}\{a, b, s, q \mid z\} = -\tfrac{N}{2} \ln(2\pi) - \tfrac{1}{2} \sum_{i=1}^{N} \{[\tilde{z}(i \mid i - 1)]^2 / \mathrm{Var}[\tilde{z}(i \mid i - 1)]$$

$$+ \ln \mathrm{Var}[\tilde{z}(i \mid i - 1)]\} + m(q)\ln(\lambda)$$
$$+ [N - m(q)]\ln(1 - \lambda) \tag{9-40}$$

The final expression for $\mathcal{M}\{a, b, s, q \mid z\}$ is obtained from Eqs. (9-37) and (9-40), and the fact (due to the first two properties of the innovations that were discussed above) that

$$\tfrac{1}{2} \ln \mid \Omega \mid = \tfrac{1}{2} \sum_{i=1}^{N} \ln \mathrm{Var}[\tilde{z}(i \mid i - 1)] ; \tag{9-41}$$

i.e.,

$$\mathcal{M}\{\mathbf{a}, \mathbf{b}, \mathbf{s}, \mathbf{q} \mid \mathbf{z}\} = -\frac{3}{2}\frac{N}{2}\ln(2\pi) - \frac{1}{2}\sum_{i=1}^{N}[\tilde{z}(i \mid i-1)]^2 / \mathrm{Var}[\tilde{z}(i \mid i-1)]$$

$$-\frac{N}{2}\ln v_n v_r v_B + m(q)\ln(\lambda) + [N - m(q)]\ln(1-\lambda) \quad (9\text{-}42)$$

Observe that to compute $\tilde{z}(i \mid i-1)$ we must run the innovations filter, which in turn requires the running of the Kalman predictor. Of course, these quantities are calculated recursively; hence, the Kalman predictor and innovations process are coupled into the calculation of $\mathcal{M}$.

9.2.5   Gradients of $\mathcal{L}$ and $\mathcal{M}$

The gradients of $\mathcal{L}$ and $\mathcal{M}$ required massive amounts of calculations when the convolutional model was used. They are much easier to compute when the state-variable model is used.

9.2.5.1   Gradients of $\mathcal{L}$

The reader should review the 9 step algorithm, given in the Chapter 7 section entitled "Gradients of $\mathcal{L}$ with Respect to $\mathbf{a}$ and $\mathbf{b}$," as well as the discussions following the algorithm to see that the 9 step algorithm does not require the solution of any $N \times N$ equations, but it does require $u^{MV}(k \mid N)$ as an approximation to $u(k)$. As summarized in Table 9-1 (see Footnote b), this latter calculation requires on the order of N^3 flops when batch processing is used. In principle, it is therefore unnecessary to change any of the steps in the 9 step algorithm. All that is necessary is to compute $u^{MV}(k \mid N)$ using the recursive algorithm described in this chapter. This reduces the number of flops from $O(N^3)$ to $O(2n^3N)$. The latter quantity is insignificant compared to the former quantity.

Steps 2 through 6 of the 9 step algorithm require calculations of the wavelet's impulse response as well as derivatives of the impulse response with respect to the a- and b- parameters of the wavelet's ARMA model. These calculations are easily performed using a finite-difference model for the ARMA wavelet. They can also be performed using the equivalent state-variable model; however, there is no computational advantage to doing that; hence, we leave details on how to do that to the reader (see, e.g., Mendel, Chi, and Hampson, 1985).

The reader should next review the section in Chapter 7 entitled "Derivatives of $\mathcal{L}$ with Respect to Variances." The 4 step algorithm for computing the derivatives of $\mathcal{L}$ with respect to v_B and v_n, again requires $u^{MV}(k \mid N)$; hence, this derivatives of $\mathcal{L}$ with respect to v_B and v_n, again requires $u^{MV}(k \mid N)$; hence, this algorithm does not need to be modified. All that is necessary is to compute $u^{MV}(k \mid N)$ using the recursive algorithm described in this chapter. Recall that it will require $O(2n^3N)$ flops.

9.2.5.2 Gradients of $\mathcal{M}$

A review of the 12 step algorithm for calculating the gradients of $\mathcal{M}$ with respect to the **a**- and **b**- parameters reveals that steps 10 and 11 are the computationally intensive ones; they require solutions of $(2n + 1)$ N×N linear equations, which requires $O[(2n + 1) N^3]$ flops. Recursive processing can be used to reduce this count; but, it requires a completely different algorithm, one that begins with the expression for $\mathcal{M}\{a, b, s, q \mid z\}$ that is given in Eq. (9-42).

Consider $\partial \mathcal{M}\{a, b, s, q \mid z\}/\partial a_j$, for example:

$$\partial \mathcal{M}\{a, b, s, q \mid z\}/\partial a_j$$

$$= -\sum_{i=1}^{N} \{\tilde{z}(i \mid i - 1)/\text{Var}[\tilde{z}(i \mid i - 1)]\}\partial\tilde{z}(i \mid i - 1)/\partial a_j$$

$$+ \tfrac{1}{2} \sum_{i=1}^{N}\{\tilde{z}(i \mid i - 1)/\text{Var}[\tilde{z}(i \mid i - 1)]\}^2\partial\text{Var}[\tilde{z}(i \mid i - 1)]/\partial a_j \qquad (9\text{-}43)$$

In order to calculate this quantity we need $\partial\tilde{z}(i \mid i - 1)/\partial a_j$ and $\partial\text{Var}[\tilde{z}(i \mid i - 1)]/\partial a_j$. These are obtained by taking the partial derivative of the innovations process and the Kalman predictor [whose equations are given as (9-22) - (9-28)], with respect to a_j. Doing this we obtain another linear system of equations for each a_j, known as the *Kalman Predictor/ Innovations Sensitivity System*. Just as the Kalman Predictor requires $O(2n^3N)$ flops, so does the Kalman Predictor/ Innovations Sensitivity System. Because there are a total of n a-parameters and n b-parameters, we will need 2n sensitivity systems. A close examination of the structure of these sensitivity systems reveals that they are driven by a Kalman predictor (for detailed structures of these systems see, e.g., Mendel, 1983); hence, the calculations of $\partial \mathcal{M}\{a, b, s, q \mid z\}/\partial a_j$ and $\partial \mathcal{M}\{a, b, s, q \mid z\}/\partial b_j$ for j = 1, 2, ... , n requires $O[(2n + 1)2n^3N]$ flops. Compared to $O[(2n + 1)N^3]$, $O[(2n + 1)2n^3N]$ is a very small number of flops; hence, there is a tremendous advantage to computing $\partial \mathcal{M}\{a, b, s, q \mid z\}/\partial a_j$ and $\partial \mathcal{M}\{a, b, s, q \mid z\}/\partial b_j$ by recursive techniques. Of course, it is important to keep the order of the ARMA model, n, as small as possible.

The same is true for calculating the derivatives of $\mathcal{M}$ with respect to v_B and v_n. For example, $\partial \mathcal{M}\{a, b, s, q \mid z\}/\partial v_B$ looks almost the same as $\partial \mathcal{M}\{a, b, s, q \mid z\}/\partial a_j$ in Eq. (9-43), except that all partial derivatives are with respect to v_B, and, there is one additional term due to the term $-\frac{N}{2} \ln v_n v_r v_B$ which appears in $\mathcal{M}$.

9.2.6   Pseudo-Hessians of $\mathcal{L}$ and $\mathcal{M}$

Recall that the Marquardt-Levenberg algorithm not only uses the gradient, but it also uses the Hessian matrix. The latter is much too complicated to calculate; hence, it is approximated by the pseudo-Hessian in which higher-order derivatives are neglected.

9.2.6.1 Pseudo-Hessian of $\mathcal{L}$

The reader should refer to the Chapter 7 section, entitled "Pseudo-Hessian of $\mathcal{L}$ with respect to **a** and **b**," to recall that $\partial^2 \mathcal{L}/\partial a_i \partial a_j$ and $\partial^2 \mathcal{L}/\partial b_i \partial b_j$ can be approximated without having to solve any N×N equations; hence, there is no need to calculate these pseudo-Hessians in a new way.

The same is true for the calculations of $\partial^2 \mathcal{L}/\partial v_B{}^2$ and $\partial^2 \mathcal{L}/\partial v_n{}^2$.

9.2.6.2 Pseudo-Hessian of $\mathcal{M}$

Calculating the pseudo-Hessians of $\mathcal{M}$ by batch processing requires an incredibly large amount of computation. These calculations can be reduced significantly when they are carried out using recursive processing.

The Hessian of $\mathcal{M}$ with respect to **a** is the n×n matrix $\{\partial^2 \mathcal{M}/\partial a_i \partial a_j\}$. The second derivative, $\partial^2 \mathcal{M}/\partial a_i \partial a_j$, can be obtained by differentiating Eq. (9-43) with respect to a_i. The result is a very complicated looking expression involving not ony first derivatives of $\tilde{z}(i \mid i - 1)$ and $\text{Var}[\tilde{z}(i \mid i - 1)]$, but also second derivatives of these quantities. The pseudo-Hessian of $\mathcal{M}$ is obtained by dropping all second derivative terms. Justification for this approximation is that as the maximum-likelihood estimates of the wavelet's parameters approach their true values, the expected value of the dropped terms go to zero (Gupta and Mehra, 1974).

Each element in the pseudo-Hessian matrix, given by the approximation to $\partial^2 \mathcal{M}/\partial a_i \partial a_j$, requires O(N) flops. Due to the symmetric nature of matrix

$\{\partial^2 \mathcal{M}/\partial a_i \partial a_j\}$, its calculation will require $O[n(n - 1)N/2]$ flops. Hence, the calculations of $\{\partial^2 \mathcal{M}/\partial a_i \partial a_j\}$ and $\{\partial^2 \mathcal{M}/\partial b_i \partial b_j\}$ will require $O[n(n - 1)N]$ flops. This is a vastly smaller number of computations than the $O[n(n - 1)N^3]$ flops that are required when batch processing is used.

No new method is needed to calculate the second derivatives of M with respect to v_B and v_n, because these second derivatives are given by the very simple formulas in (7-170) and (7-171), respectively.

9.2.7   Computational Requirements for Recursive Processing

Table 9-2 summarizes the gross computational requirements for each component of the Figure 9-1 Block Component Method. As in Table 9-1, this is done for both $\mathcal{L}$ and $\mathcal{M}$.

Assuming α iterations in the Adaptive Detection block, β iterations in the Update Wavelet Parameters block, and γ iterations in the Update Standard Deviations block, and one calculation of $\mathcal{L}$ or $\mathcal{M}$ per iteration within each block, the total gross numbers of flops when we use $\mathcal{M}$ or $\mathcal{L}$, denoted $\text{Flops}_R(\mathcal{M})$ and $\text{Flops}_R(\mathcal{L})$, where the subscript 'R' denotes 'recursive' calculations, are

$$\begin{aligned}
\text{Flops}_R(\mathcal{M}) &= 4n^3N + \alpha(4n^3N) + \beta[(2n + 1)2n^3N] + \gamma(6n^3N) + 2n^3N \\
&= 6n^3N + 4n^3N\alpha + (2n + 1)2n^3N\beta + 6n^3N\gamma
\end{aligned} \tag{9-44}$$

and

$$\begin{aligned}
\text{Flops}_R(\mathcal{L}) &= 4n^3N + \alpha(4n^3N) + \beta(2n^3N) + \gamma(2n^3N) + 2n^3N \\
&= 6n^3N + 4n^3N\alpha + 2n^3N\beta + 2n^3N\gamma
\end{aligned} \tag{9-45}$$

Observe that the major difference between $\text{Flops}_R(\mathcal{M})$ and $\text{Flops}_R(\mathcal{L})$ is in the factor $4n^4N$ in $\text{Flops}_R(\mathcal{M})$, which is associated with the calculation of the $2n$ gradients of $\mathcal{M}$ with respect to the wavelet's parameters.

To illustrate these results, we use the example presented in the first section of this chapter, for which: $N = 1,000$, $n = 6$ (i.e., we have a 6th-order ARMA wavelet), and, $\alpha = \beta = \gamma = 1$. If 1 flop requires 10^{-6} sec. of computation time, then $\text{Flops}_R(\mathcal{M})$ requires 9.072 sec., whereas $\text{Flops}_R(\mathcal{L})$ requires 3.024 sec. Table 9-3 compares the results for this example, for both batch and recursive processing. By clever coding of the recursive equations, the computation times for recursive processing can be reduced much further, and, if parallel computation is used, then all computations will be well under 1 sec. Consequently, recursive processing is quite practical, and is absolutely needed in

Table 9-2. Computational Requirements for Figure 9-1 Block Component Method: Recursive Calculations.

COMPONENT	EQUATIONS	
Initialize Parameters		
MVD (needed for Threshold Detector)	(9-18)-(9-28)[b]	
Threshold Detector	(7-62) to (7-64)	
Adaptive Detection		
SMLR Detector	(7-93)&(7-95)[c]	
λ-Update	(7-184)	
Update Wavelet Parameters	$\mathcal{M}$	$\mathcal{L}$
Gradients	Kalman predictor/innovations sensitivity systems	9 step algorithm[d]
Pseudo-Hessians	Mendel (1983)	(7-166) to (7-169)
Marquardt-Levenberg[e]	(4-8)&(4-9)	(4-8)&(4-9)
Update Standard Deviations	$\mathcal{M}$	$\mathcal{L}$
Gradients	5 step algorithm	4 step algorithm[d]
Pseudo-Hessians	(7-170)&(7-171)	(7-174)&(7-175)
Marquardt-Levenberg	(4-8)&(4-9)	(4-8)&(4-9)
Stop?[f]	$\mathcal{M}$	$\mathcal{L}$
	(9-42)	(5-22) or (9-35)
Update Random Parameters[g]	(9-18),(9-20),(9-22)-(9-28)	

a. Only $O(N^3)$ calculations are listed. A blank indicates $O(n^2N)$ or $O(nN)$ calculations.
b. The intensive computations are in Eqs. (9-21) and (9-25).
c. See also Eqs. (7-98) and (7-99). MVD is needed by the SMLR detector.
d. Requires (at least) one calculation of $u^{MV}(k \mid N)$ that is used in *both* gradient and pseudo-Hessian calculations.

Table 9-2 (continued)

FLOPS[a]	
$4n^3N$	
--	
$4n^3N$	
--	
$\mathcal{M}$	$\mathcal{L}$
$(2n + 1)2n^3N$	$2n^3N$
--	--
--	--
$\mathcal{M}$	$\mathcal{L}$
$(2+1)2n^3N$	$2n^3N$
--	--
--	--
$\mathcal{M}$	$\mathcal{L}$
--	--
$2n^3N$	

e. $(H_{a,i} + D_{a,i})^{-1}$ and $(H_{b,i} + D_{b,i})^{-1}$ each require $O(n^3)$ flops.

f. If this BCM is programmed for an interactive environment, in which the user chooses how many iterations are spent in each block, then $\mathcal{M}$ and $\mathcal{L}$ are usually computed for each iteration in a block and are displayed to the user, so that the user can decide whether or not it is worthwhile to iterate further in the block.

g. The intensive computation is in Eq. (9-25).

order to make MLD a practical tool.

Finally, in Chapter 7 we mentioned that storage of large N×N matrices, such as Ω, can be another problem for batch processing. When the recursive processing techniques of this chapter are used, then we only need to store N n×n matrices, such as $P(k + 1 \mid k)$, or N n×1 vectors, such as $K(k)$. Using careful programming techniques, some of these matrices and vectors can be made to use the same storage.

Table 9-3. Numerical comparisons: N = 1,000, n = 6, $\alpha = \beta = \gamma = 1$, and 1 flop in 10^{-6} sec.

Processing	Flops($\mathcal{M}$) in:	Flops($\mathcal{L}$) in:
Batch	16.11 hr.	3.89 hr.
Recursive	9.072 sec.	3.024 sec.

9.3 Summary

While it is easy to explain the entire theory of maximum-likelihood deconvolution using the convolutional model and batch processing, in practice, using today's traditional computers, it is impossible to compute using batch processing. It is practical to compute using the recursive processing formulas which are stated in this chapter. These formulas make use of the recursive nature of the ARMA wavelet model, whereas the batch formulas do not.

In all of our work we have separated amplitude restoration and event detection, justifiably so, because of the Separation Principle. Another approach, which does not rely on the Separation Principle, is to reconstruct the reflectivity in a maximum a posteriori framework. In this approach one tries to learn the a posteriori probabilities that q(k) = 0 or 1. Hopefully, at each time point these probabilities will converge to zero or unity. Papers which take this point of view are: Mahalanabis, et. al. (1982), Katayama and Tanaka (1986), and Wang and Dai (1988).

Finally, if the wavelet changes over the time window of N measurements (i.e., it is time-varying), then shorter data windows must be used, over which the wavelet can be thought of as time-invariant; or, more real-time wavelet identification techniques must be used, in order to track the wavelet. The latter approaches are usually referred to as adaptive deconvolution techniques (e.g., Kollias and Halkias, 1985). They must, of course, be accompanied by more real-time detection and amplitude restoration techniques.

References

Akaike H (1974) A new look at statistical model identification. IEEE Trans Auto Cont AC19:716-723

Anderson BDO, Moore JB (1979) *Optimal Filtering*. New Jersey: Prentice-Hall Inc

Berkhout AJ (1974) Related properties of minimum-phase and zero-phase time functions. Geophys Prosp 22:683-709

Blass WE, Halsey GW (1981) *Deconvolution of Absorption Spectra*. New York: Academic Press

Breiman L (1968) *Probability*. Massachusetts: Wesley

Cappello PR, et al (eds) (1984) *VLSI Signal Processing*. New York: IEEE Press

Chi C-Y (1983) Single-channel and multichannel deconvolution. PhD Thesis, Univ of Southern California, Los Angeles CA

Chi C-Y (1987) A fast maximum-likelihood estimation and detection algorithm for Bernoulli-Gaussian processes. IEEE Trans Acoust Speech Sig Proc ASSP-35:1636-1639

Chi C-Y, Mendel JM (1984a) Improved maximum-likelihood detection and estimation of Bernoulli Gaussian processes. IEEE Trans Info Theory IT-30:429-435

Chi C-Y, Mendel JM (1984b) Performance of minimum-variance deconvolution filter. IEEE Trans Acoust Speech Sig Proc ASSP-32:1145-1153

Chi C-Y, Goutsias J, Mendel JM (1985) A fast maximum-likelihood estimation and detection algorithm for Bernoulli-Gaussian processes. Presented at 1985 ICASSP, Tampa FL

Chi C-Y, Mendel JM, Hampson D (1984) A computationally-fast approach to maximum-likelihood deconvolution. Geophysics 49:550-565

Chu JJ, Wang W (1985) Parameter estimation and identification of nonminimum phase linear stochastic systems. In: *Preprints of IFAC Identification and Systems Parameter Estimation Conf*, York, England, pp. 991-996

Claerbout JF (1978) Minimum information deconvolution. Tech Report, Exploration Project 13, Stanford Univ, Stanford CA

Claerbout JF (1979) Parsimonious deconvolution. Tech Report, Exploration Project 15, Stanford Univ, Stanford CA

Deeming TJ (1984) Why minimum entropy deconvolution doesn't work. Presented at SEG Workshop on Deconvolution, Vail CO

Demoment G, Reynaud R (1985) Fast minimum variance deconvolution. IEEE Trans Acoust Speech Sig Proc ASSP-33:1324-1326

Demoment G, Reynaud R, Herment A (1984) Range resolution improvement by a fast deconvolution method. Ultrasonic Imaging 6:435-451

Deng Z-I (1985) White noise filter and smoother with application to seismic data deconvolution. Proc IFAC Ident Sys Param Est, pp. 621-624

Dongarra JJ, Bunch JR, Moler CB, Stewart GW (1979) *LINPACK User's Guide*. Pennsylvania: SIAM

Edwards AWF (1972) *Likelihood*. London: Cambridge Univ Press

Fisher RA (1922) On the mathematical foundations of theoretical statistics. Philos Trans Royal Soc London Ser A 222:309-368

Fisher RA (1925) Theory of statistical estimation. Cambridge Philos Soc Proc 22:700-725

Fox RL (1971) *Optimization Methods for Engineering Design*. Massachusetts: Wesley

Gabel RA, Roberts RA (1980) *Signals and Linear Systems*, 2nd ed. New York: Wiley

Giannakis GB, Mendel JM (1987) Entropy interpretation of maximum likelihood deconvolution. Geophysics 52:1621-1630

Giannakis GB, Mendel JM (1989) Identification of non-minimum phase systems using higher-order statistics. IEEE Trans Acoust Speech Sig Proc ASSP-37: 360-377

Giannakis GB, Mendel JM, Zhao X (1987) A fast prediction-error detector for estimating sparse-spike sequences. Presented at IEEE ICASSP, Dallas TX; also in IEEE Trans Geosci Rem Sens GE-27 (1989):344-351

Godfrey R (1978) An information theory approach to deconvolution. Tech Report Exploration Project 13, Stanford Univ, Stanford CA

Godfrey R, Rocca F (1981) Zero memory non-linear deconvolution. Geophys Pros 29:189-228

Goutsias J, Mendel JM (1986) Maximum-likelihood deconvolution: an optimization theory perspective. Geophysics 51:1206-1220

Goussard Y, Demoment G (1987a) Recursive deconvolution of Bernoulli-Gaussian processes using a MA representation. Private communication

Goussard Y, Demoment G (1987b) Fast recursive detection-estimation of Bernoulli-Gaussian processes. Modelisation-Identification, private communication

Gray W (1979) Variable norm deconvolution. PhD Thesis, Stanford Univ, Stanford CA

Gupta NK, Mehra RK (1974) Computational aspects of maximum likelihood estimation and reduction of sensitivity function calculations. IEEE Trans Auto Cont AC-19:774-783

Hampson D, Russell BH (1983) The performance of maximum-likelihood deconvolution on real seismic data. Presented at the CSEG Mtg, Calgary, Canada

Hsueh A-C (1983) State variable modeling and recursive processing of 1-D and 2-D non-causal systems. PhD Thesis, Univ of Southern California, Los Angeles CA

Hsueh A-C, Mendel JM (1985) Minimum-variance and maximum-likelihood deconvolution for noncausal channel models. IEEE Trans Geosci Rem Sens GE-23:797-808

Katayama T, Tanaka M (1986) Detection and estimation of a Bernoulli-Gauss process for linear discrete-time systems. Intl J Systems Sci 17:687-702

Kollias SD, Halkias CC (1985) An instrumental variable approach to minimum-variance seismic deconvolution. IEEE Trans Geosci Rem Sens GE-23:778-788

Kollias SD, Manolakis D, Halkias CC (1985) ARMA identification methods in the deconvolution of seismic signals. Intl J Mod Simul 5:85-88

Kormylo J (1979) Maximum-likelihood seismic deconvolution. PhD Thesis, Univ of Southern California, Los Angeles CA

Kormylo J, Mendel JM (1980) Simultaneous spherical divergence correction and optimal deconvolution. IEEE Trans Geosci Rem Sens 18:273-280

Kormylo J, Mendel JM (1982) Maximum-likelihood detection and estimation of Bernoulli-Gaussian processes. IEEE Trans Info Theory IT-28:482-488

Kormylo J, Mendel JM (1983a) Identifiability of non-minimum phase linear stochastic systems. IEEE Trans Auto Cont AC-28:1081-1090

Kormylo J, Mendel JM (1983b) Maximum-likelihood deconvolution. IEEE Trans Geosci Rem Sens GE-21:72-82

Kung SY (1978) A new identification and model reduction algorithm via singular value decomposition. Presented at 12th Annual Asilomar Conf on Circuits, Systems, and Computers, Pacific Grove CA

Kung SY (1987) *VLSI Array Processors*. New Jersey: Prentice-Hall Inc

Kung SY, et al (eds) (1986) *VLSI Signal Processing, II*. New York: IEEE Press

Kwakernaak H (1980) Estimation of pulse heights and arrival times. Automatica 16:367-377

Lainiotis DG, Katsikas SK, Likothanasis SD (1988a) Optimal seismic deconvolution. Signal Processing 15:375-404

Lainiotis DG, Katsikas SK, Likothanasis SD (1988b) Adaptive deconvolution of seismic signals-performance, computational analysis, parallelism. IEEE Trans Acoust Speech Sig Proc ASSP-36:1715-1734

Lines LR, Clayton RW (1977) A new approach to Vibroseis deconvolution. Geophys Pros 25:417-433

Ljung L (1987) *System Identification: Theory for the User*. New Jersey: Prentice-Hall Inc

Mahalanabis AK, Prasad S, Mohandas KP (1982) Recursive decision directed estimation of reflection coefficients for seismic data deconvolution. Automatica 18:721-726

Mansuripur M (1987) *Introduction to Information Theory*. New Jersey: Prentice-Hall Inc

Marquardt DW (1963) An algorithm for least-squares estimation of nonlinear parameters. J Soc Indust Appl Math 11:431-441

Marschall R (1977) Wavelet processing by means of recursive filters. Presented at 47th Annual Intl Mtg of the SEG, Calgary, Canada

Meditch JS (1969) *Stochastic Optimal Linear Estimation and Control*. New York: McGraw-Hill

Melsa JL, Cohn DL (1978) *Decision and Estimation Theory*. New York: McGraw-Hill

Mendel JM (1977) White noise estimators for seismic data processing in oil exploration. IEEE Trans Auto Cont AC-22:694-706

Mendel JM (1981) Minimum-variance deconvolution. IEEE Trans Geosci Rem Sens GE-19:161-171

Mendel JM (1983) *Optimal Seismic Deconvolution: An Estimation-Based Approach*. New York: Academic Press

Mendel JM (1984) Simultaneous correction for divergence and deconvolution without changing industrial practice. Geophysics 49:584-585

Mendel JM (1985) How to include prespecified horizons into minimum-variance deconvolution and maximum-likelihood deconvolution. Geophysics 50:1510-1512

Mendel JM (1986) Some modeling problems in reflection seismology. IEEE ASSP Mag 3:4-17

Mendel JM (1987a) *Lessons in Digital Estimation Theory*. New Jersey: Prentice-Hall Inc

Mendel JM (1987b) *Kalman Filtering and Other Digital Estimation Techniques: Study Guide*. New York: IEEE, Individual Learning Package

Mendel JM, Kormylo J (1978) Single channel white-noise estimators for deconvolution. Geophysics 43:102-124

Mendel JM, Kormylo J, Aminzadeh F, Lee JS, Habibi-Ashrafi F (1981) A novel approach to seismic signal processing and modeling. Geophysics 46:1398-1414

Nahi NE (1969) *Estimation Theory and Applications*. New York: Wiley

Ooe M, Ulrych TJ (1979) Minimum entropy deconvolution with an exponential transformation. Geophys Prosp 27:458-473

Papoulis A (1984) *Probability, Random Variables, and Stochastic Processes.* New York: McGraw-Hill

Popoli RF, Mendel JM (1989) Heuristically constrained estimation for intelligent signal processing.   Presented at the 1987 IEEE Conf on Decision and Control, Los Angeles CA; also in: Aminzadeh F, Simaan, M (eds) *Artificial Intelligence and Expert Systems in Petroleum Exploration.* Massachusetts: JAI Press

Robinson EA (1967) *Multichannel Time Series Analysis with Digital Computer Programs.* San Francisco: Holden-Day Inc

Robinson EA (1983) *Seismic Velocity Analysis and the Convolutional Model.* Massachusetts: Intl Human Resources Development Corp

Robinson EA, Treitel S (1980) *Geophysical Signal Analysis.* New Jersey: Prentice-Hall Inc

Scargle JD (1977) Absolute value optimization to estimate phase properties of stochastic time series. IEEE Trans Info Theory IT-23:140-143

Shiva M (1982) Geo-optimal deconvolution.   PhD Thesis,   Univ of Southern California, Los Angeles CA

Sorenson HW (1980) *Parameter Estimation: Principles and Problems.* New York: Dekker

Stewart GW (1973) *Introduction to Matrix Computations.* New York: Academic Press

Tugnait JK (1985) Identification of Non-minimum phase linear stochastic systems. Automatica 22:457-464

Walden AT (1985) Non-Gaussian reflectivity, entropy and deconvolution. Geophysics 50:2862-2888

Walden AT, Hosken JWJ (1986) The nature of the non-Gaussianity of primary reflection coefficients and its significance for deconvolution. Geophys Pros 34:1038-1066

Wang L-X, Dai G-Z (1988) New recursive smoothing algorithms for Bernoulli-Gaussian input sequences.   Presented at the Intl Federation on Automatic Control, Beijing, People's Republic of China

Wiggins RA (1977) Minimum entropy deconvolution. Proc Intl Symp Computer Aided Seismic Analysis and Discrmination. Falmouth MA, IEEE Computer Soc, pp. 7-14

Wiggins RA (1978) Minimum entropy deconvolution. Geoexploration 16:21-35

Zhao X, Mendel JM (1988a) Minimum-variance deconvolution using artificial neural networks. Presented at the 58th Annual Intl Mtg of the SEG, Anaheim CA

Zhao X, Mendel JM (1988b) An artificial minimum-variance estimator. Proc IEEE Intl Conf on Neural Networks, Vol II: 499-506, San Diego CA

Bibliography

Aboutajdine D, Najim M (1985) Comments on Adaptive Filter Structures for Deconvolution of Seismic Signals. IEEE Trans Geosci Rem Sens GE-23:72-73

Alam AM (1974) A sequential adaptive deconvolution algorithm based on Kalman-filter approach. Presented at 44th Annual Intl Mtg of the SEG, Dallas TX

Alam AM, Sicking CJ (1981) Recursive removal and estimation of minimum-phase wavelet. Geophysics 46:1379-1391

Alcazar-Fernandez JJ, Casar-Corredera R, Garcia-Gomez R (1986) L_1-norm noisy Tauberian deconvolution. Presented at 1986 ICASSP, Tokyo, Japan

Arya VK, Aggarwal JG (1982) *Deconvolution of Seismic Data*, Dowden: Benchmark Papers in Electrical Engineering and Computer Science; Pennsylvania: Hutchinson Ross

AuYeung C, Mersereau RM, Schafer RW (1986) Maximum entropy deconvolution. Presented at 1986 ICASSP, Tokyo, Japan

Backus M (1959) Water reverberations, their nature and elimination. Geophysics 24:233-262

Bardan V (1977) Comments on dynamic predictive deconvolution. Geophys Prosp 25:569-572

Bayless JW, Brigham EO (1970) Application of the Kalman filter to continuous signal restoration. Geophysics 35:2-23

Bednar JB (1983) Applications of median filtering to deconvolution, pulse estimation, and statistical editing of seismic data. Geophysics 48:1598-1610

Bellini S (1986) Bussgang Techniques for Blind Equalization. Presented at GLOBECOM '86, Houston TX

Benveniste A, Goursat M, Ruget G (1980) Robust identification of a non-minimum phase system: blind adjustment of linear equalizer in data communication. IEEE Trans Auto Cont AC-25:385-398

Berkhout AJ (1974) Related properties of minimum-phase and zero-phase time functions. Geophys Prosp 22:683-709

Berkhout AJ (1977) Least squares inverse filtering and wavelet deconvolution. Geophysics 42:1369-1383

Berman M (1978) A deconvolution scheme. Math Biosc 40:319-323

Bickel SH (1982) The effects of noise on minimum-phase vibroseis deconvolution. Geophysics 47:1174-1184

Bickel SH, Natarajan RR (1985) Plane-wave Q deconvolution. Geophysics 50:1426-1439

Blass WE, Halsey GW (1981) *Deconvolution of Absorption Spectra*. New York: Academic Press

Boland FM, Doyle T (1982) Deconvolution in real time of noisy signals. Presented at 1982 ICASSP, Paris, France, pp. 1853-1857

Brillinger DR (1988) Some statistical methods for random process data from seismology and neurophysiology. The Annals of Statistics 16:1-54

Brown RL, McElhattan W, Santiago DJ (1988) Wavelet estimation: an interpretive approach. Geophysics: The Leading Edge of Exploration 7:16-19

Buhl P, Stoffa PL, Bryan GM (1974) Application of homomorphic deconvolution to shallow-water marine seismology. Geophysics 39:401-426

Bunks C, Preis D (1980) Minimax time-domain deconvolution for transversal filter equalizers. Presented at 1980 ICASSP, Denver CO:943-946

Butkus B (1975) Homomorphic filtering, theory and practice. Geophys Prosp 23:712

Carlini A, Storer P (1987) Experiences in signal deconvolution for seismic stratigraphy. In: Worthington MH (ed) *Deconvolution and Inversion.* Oxford England: Blackwell Scientific Publ, pp. 52-66

Cavin III RK, McCormack MD, Verm RW, Holyoak JN (1977) Estimation theory approach to non-stationary deconvolution. Presented at 47th Annual Intl Mtg of the SEG, Alberta, Canada

Cavin III RK, McCormack MD, Verm RW, Holyoak JN (1978) The influence of statistical parameter estimates on Kalman and Wiener deconvolution. Dept of EE, Texas A&M University, Houston TX, private communication

Chang JY, Backus MM, (1981) Emergent angle dependent deconvolution. Presented at the 51st Annual Intl Mtg of the SEG, Los Angeles CA

Chi C-Y (1983) Single-channel and multichannel deconvolution. PhD Thesis, Univ of Southern California, Los Angeles CA

Chi C-Y (1987) A further analysis for the minimum-variance deconvolution filter performance. IEEE Trans Acoust Speech Sig Proc ASSP-35:888-889

Chi C-Y (1987) A fast maximum-likelihood estimation and detection algorithm for Bernoulli-Gaussian processes. IEEE Trans Acoust Speech Sig Proc ASSP-35:1636-1639

Chi C-Y, Goutsias J, Mendel JM (1985) A fast maximum-likelihood estimation and detection algorithm for Bernoulli-Gaussian processes. Presented at 1985 ICASSP, Tampa FL

Chi C-Y, Mendel JM (1983) A fast approach to identification using deconvolution. Presented at the IEEE Conf on Decision and Control, San Antonio TX

Chi C-Y, Mendel JM (1984) Improved maximum-likelihood detection and estimation of Bernoulli Gaussian processes. IEEE Trans Info Theory IT-30:429-435

Chi C-Y, Mendel JM (1984) Performance of minimum-variance deconvolution filter. IEEE Trans Acoust Speech Sig Proc ASSP-32:1145-1153

Chi C-Y, Mendel JM (1985) Multichannel maximum-likelihood deconvolution. Presented at 55th Intl Mtg of the SEG, Washington DC

Chi C-Y, Mendel JM (1985) Viterbi algorithm detector for Bernoulli-Gaussian processes. IEEE Trans Acoust Speech Sig Proc ASSP-33:511-519

Chi C-Y, Mendel JM, Hampson D (1984) A computationally-fast approach to maximum-likelihood deconvolution. Geophysics 49:550-565

Claerbout JF (1978) Minimum information deconvolution. Tech Report, Exploration Project 13, Stanford Univ, Stanford CA

Claerbout JF (1979) Parsimonious deconvolution. Tech Report, Exploration Project 15, Stanford Univ, Stanford CA

Claerbout JF (1986) Simultaneous pre-normal moveout and post-normal moveout deconvolution. Geophysics 51:1341-1354

Clarke GKC (1968) Time-varying deconvolution filters. Geophysics 33:936-944

Clayton RW, Ulrych TJ (1977) A restoration method for impulsive functions. IEEE Trans Info Theory IT-23:262-264

Clayton RW, Wiggins RA (1977) Source shape estimation and deconvolution of teleseismic bodywaves. J Royal Astron Soc Gephys 4-F:151-177

Commenges D (1984) The deconvolution problem: fast algorithms including the preconditioned conjugate-gradient to compute a MAP estimator.   IEEE Trans Auto Cont AC-29:229-243

Connelly D, Hart D, Dragoset B, Hargreaves N, Larner K (1987) The 'model-based' approach to wavelet processing. In: Worthington MH (ed) *Deconvolution and Inversion.* Oxford, England: Blackwell Scientific Publ, pp. 67-89

Crump ND (1974) A Kalman filter approach to the deconvolution of seismic signals. Geophysics 39:1-13

Crump ND (1975) Techniques for the deconvolution of seismic signals. Proc IEEE Conf on Decision and Control, Houston TX:2-8

Dai G-Z, Mendel JM (1986) General problems of minimum-variance recursive waveshaping.   IEEE Trans Acoust Speech Sig Proc ASSP-34:616-618

Dai G-Z, Mendel JM (1986) A straightforward and unified approach to the derivation of minimum-variance deconvolution algorithms.   IEEE Trans Auto Cont AC-31:80-83

Deeming TJ (1984) Why minimum entropy deconvolution doesn't work. Presented at SEG Workshop on Deconvolution, Vail CO

Deeming TJ, Taylor HL (1981) Use of an improved l_1 norm minimization algorithm for wavelet processing of seismic sections. Presented at 51st Annual Intl Mtg of the SEG, Los Angeles CA

Demoment G, Reynaud R (1985) Fast minimum variance deconvolution. IEEE Trans Acoust Speech Sig Proc ASSP-33:1324-1326

Demoment G, Reynaud R, Herment A (1983) Fast minimum variance deconvolution with applications to ultrasound tissue characterization. Lab des Signaux et Systemes, France, Internal Report No LSS/061/1983

Demoment G, Reynaud R, Herment A (1984) Range resolution improvement by a fast deconvolution method. Ultrasonic Imaging 6:435-451

Deng Z-I (1985) White noise filter and smoother with application to seismic data deconvolution. Proc IFAC Identification and System Parameter Estimation, pp. 621-624

D'Mello MR (1974) Adaptive filtering and smoothing in deconvolution for seismic processes. MS Thesis, Oklahoma State Univ, Stillwater OK

Donoho D (1981) On minimum entropy deconvolution. In: Findley DF (ed) *Applied Time Series Analysis II,* New York: Academic Press, pp. 565-608

Douze EJ (1971) Prediction error filters, white noise and orthogonal coordinates. Geophys Prosp 19: 253-264

Egan MS, Craft KL, Reed R (1988) Dip-dependent deconvolution. Presented at the 58th Annual Intl Meeting of the SEG, Anaheim CA

Ekstrom MP (1973) A spectral characterization of the ill conditioning in numerical deconvolution. IEEE Trans Audio Electroacoust AU-21: 344-348

Eisenstein BA, Cerrato LR (1976) Statistical deconvolution. J Franklin Inst 302: 147-157

Ford WT, Hearne JH (1966) Least-squares inverse filtering. Geophysics 31:917-926

Foster MR, Hicks WG, Nipper JT (1962) Optimum inverse filters which shorten the spacing of velocity logs. Geophysics 27:317-326

Foster MR, Sengbush R, Watson R (1968) Use of Monte Carlo techniques in optimum design of the deconvolution process. Geophysics 33:945-949

Fourmann JM (1980) Signature processing-review and comparison of some processing principles. Presented at 42nd Annual EAEG Mtg, Istanbul, Turkey

Fryer GJ, Odegard ME, Sutton GH (1975) Deconvolution and spectral estimation using final prediction error. Geophysics 40:411-425

Fullagar PK (1985) Spike recovery deconvolution. In: Fitch AA (ed) *Developments in Geophys Explor Methods*, vol. 6. London: Elsevier

Gabrielson T (1983) Impulse response of the ocean floor: nonlinear techniques for measurement enhancement. Naval Air Development Center, Report No NADC-82253-30, Warminister PA

Gans WL (1983) The measurement and deconvolution of time jitter in equivalent-time waveform samplers. IEEE Trans Inst Meas IM-32:126-133

George CF, Smith HW, Bostick FX (1962) The application of inverse convolution techniques to improve signal response of recorded geophysical data. Proc IRE 50:2313-2319

Giannakis GB, Mendel JM (1987) Entropy interpretation of maximum likelihood deconvolution. Geophysics 52:1621-1630

Giannakis GB, Mendel JM, Zhao X (1987) A fast prediction-error detector for estimating sparse-spike sequences. Presented at IEEE ICASSP, Dallas TX; also, (1989) IEEE Trans Geosci Rem Sens GE-27:344-351

Godfrey R (1978) An information theory approach to deconvolution. Tech Report, Exploration Project, vol. 13, Stanford Univ, Stanford CA

Godfrey R, Rocca F (1981) Zero memory non-linear deconvolution. Geophys Prosp 29:189-228

Goussard Y, Demoment G (1987a) Recursive deconvolution of Bernoulli-Gaussian processes using a MA representation, private communication

Goussard Y, Demoment G (1987b) Fast recursive detection-estimation of Bernoulli-Gaussian processes. Modelisation-Identification, private communication

Goutsias J, Mendel JM (1986) Maximum-likelihood deconvolution: an optimization theory perspective. Geophysics 51:1206-1220

Gray W (1979) Variable norm deconvolution. PhD Thesis, Stanford Univ, Stanford CA

Griffiths LJ, Smolka FR, Trembly LD (1977) Adaptive deconvolution: a new technique for processing time-varying seismic data. Geophysics 42:742-759

Halsey G, Blass WE (1977) Deconvolution of infrared spectra in real time. J Appl Opt 16:286

Hampson D, Russell BH (1983) The performance of maximum-likelihood deconvolution on real seismic data. Presented at the CSEG Mtg, Calgary, Canada

Hildebrand HA (1979) Stationary and non-stationary least-squared error filtering and smoothing of seismic data. Presented at 1979 JACC, Denver CO

Hosken JWJ, Longbottom J, Walden AT, White RE (1987) Maximum kurtosis phase and the phase of the seismic wavelet. In: Worthington MH (ed) *Deconvolution and Inversion*. Oxford, England: Blackwell Scientific Publ, pp. 38-51

Hsueh A-C (1983) State variable modeling and recursive processing of 1-D and 2-D non-causal systems. PhD Thesis, Univ of Southern California, Los Angeles CA

Hsueh A-C, Mendel JM (1985) Minimum-variance and maximum-likelihood deconvolution for noncausal channel models. IEEE Trans Geosci Rem Sens GE-23:797-808

Hubbard TP (1979) Deconvolution of surface recorded data using vertical seismic profiles. Presented at 49th Annual Intl Mtg of the SEG, New Orleans LA

Jacovitti G, Neri A, Scarano G (1987) Complex reflectivity based non-minimum phase deconvolution. In: Worthington MH (ed) *Deconvolution and Inversion*. Oxford, England: Blackwell Scientific Publ, pp. 145-161

Jansson PA (1984) *Deconvolution with Applications in Spectroscopy*. New York: Academic Press

Jin DJ, Rogers JR (1983) Homomorphic deconvolution. Geophysics 48:1014-1016

Jovanovich DB, Sumner RD, Akins-Easterlin SL (1983) Ghosting and marine signature deconvolution: a prerequisite for detailed seismic interpretation. Geophysics 48:1468-1485

Jurkevics A, Wiggins R (1984) A critique of seismic deconvolution methods. Geophysics 49:2109-2116

Katayama T, Tanaka M (1986) Detection and estimation of a Bernoulli-Gauss process for linear discrete-time systems. Int J Systems Sci 17:687-702

Kelamis PG, Chiburis EF (1988) Post-critical wavelet estimation and deconvolution. Geophys Prosp 36:504-522

Koehler F, Taner MT (1985) The use of the conjugate-gradient algorithm in the computation of predictive deconvolution operators. Geophysics 50:2752-2758

Kolbjornsen K, Tjore R, Zimmermann J (1979) Practical aspects of signature processing. Presented at 49th Annual Intl Mtg of the SEG, New Orleans LA

Kollias SD (1984) Adaptive parameter estimation and deconvolution of seismic signals. PhD Dissertation, Dept of Electrical Engineering, Natl Tech Univ of Athens, Greece

Kollias SD, Halkias CC (1985) An instrumental variable approach to minimum-variance seismic deconvolution. IEEE Trans Geosci Rem Sens GE-23:778-788

Kollias S, Panayotis F, Halkias CC (1984) Sequential deconvolution of time varying seismic signal models. Proc Intl Conf on DSP, Florence, Italy

Kollias S, Manolakis D, Halkias CC (1985) ARMA identification methods in the deconvolution of seismic signals. Intl J Mod Simul 5:85-88

Kollias SD, Foudopoulos P, Halkias CC (1988) One-dimensional seismic inversion using adaptive deconvolution. Signal Processing 14:269-285

Kormylo J (1979) Maximum-likelihood seismic deconvolution. PhD Dissertation, Dept of Electrical Engineering, Univ of Southern California, Los Angeles CA

Kormylo J, Mendel JM (1979) Applying maximum-likelihood deconvolution to well-log impedance data. Presented at the 49th Annual Intl Mtg of the SEG, New Orleans LA

Kormylo J, Mendel JM (1980) Simultaneous spherical divergence correction and optimal deconvolution. IEEE Trans Geosci Rem Sens GE-18:273-280

Kormylo J, Mendel JM (1982) Maximum-likelihood detection and estimation of Bernoulli-Gaussian processes. IEEE Trans Info Theory IT-28:482-488

Kormylo J, Mendel JM (1983) Identifiability of non-minimum phase linear stochastic systems. IEEE Trans Auto Cont AC-28:1081-1090

Kormylo J, Mendel JM (1983) Maximum-likelihood deconvolution. IEEE Trans Geosci Rem Sens GE-21:72-82

Kunetz G, Fourmann JM (1968) Efficient deconvolution of marine seismic records. Geophysics 33:412-423

Kwakernaak H (1980) Estimation of pulse heights and arrival times. Automatica 16:367-377

Lackoff MR, LeBlanc LR (1975) Frequency-domain seismic deconvolution filtering. J Acoust Soc Am 57:151-159

Lacombe N, Levy A (1978) A parametric deconvolution method: application to two bands of N_2O in the 1.9 μm region. J Mol Spectrosc 71:175

LaCoste LJB   (1982) Deconvolution by successive approximations. Geophysics 47:1724-1730

Lainiotis   DG,   Katsikas   SK,   Likothanasis   SD   (1988a)   Optimal   seismic deconvolution. Signal Processing 15:375-404

Lainiotis DG, Katsikas SK, Likothanasis SD (1988b) Adaptive deconvolution of seismic   signals-performance,   computational   analysis,   parallelism.   IEEE Trans Acoust Speech Sig Proc ASSP-36:1715-1734

Leahy R,   Jeffs B (1988) Maximally sparse optimization for array beamforming and other applications. SIPI Report 131, Univ of Southern California, Los Angeles CA

Levy S, Clowes RM (1980) Debubbling: a generalized linear inverse approach. Geophys Prosp 28:840-858

Levy S, Oldenburg DW (1982) The deconvolution of phase-shifted wavelets. Geophysics 47:1285-1294

Lii KS, Rosenblatt M (1982) Deconvolution and estimation of transfer function phase   and   coefficients   for   nongaussian   linear   processes.   The Annals of Statistics 10:1195-1206

Lindsey JP,   Patch JR (1984) Deconvolution: the   consequences   of   and alternatives   to   two   of   its   assumptions.   Presented   at   the   Seismic Deconvolution Workshop, Vail CO

Lines LR, Clayton RW   (1977) A new approach to Vibroseis deconvolution. Geophys Prosp 25:417-433

Lines LR, Ulrych TJ (1977) The old and the new in seismic deconvolution and wavelet estimation. Geophys Prosp 25:512-540

MacAdam DP (1970) Digital image restoration by constrained deconvolution. J Opt Soc Am 60:1617

Mahalanabis AK, Prasad S, Mohandas KP (1981) A fast optimal deconvolution algorithm for real seismic data using Kalman predictor model. IEEE Trans Geosci Rem Sens GE-19:216-221

Mahalanabis AK, Prasad S, Mohandas KP (1982) Recursive decision directed estimation of reflection coefficients for seismic data deconvolution. Automatica 18:721-726

Mahalanabis   AK,   Prasad   S,   Mohandas   KP   (1983)   Deconvolution   of nonstationary seismic data using adaptive lattice filters. IEEE Trans Acoust Speech Sig Proc ASSP-31:591-598

Mahalanabis AK, Prasad S, Mohandas KP (1983) On the application of the fast Kalman algorithm to adaptive deconvolution of seismic data. IEEE Trans Geosci Rem Sens GE-21:426-433

Manolakis D, Carayannis G, Kalouptsidis N, Halkias CC (1983) Fast design of waveshaping filters for seismic signal deconvolution. Neuvieme Colloque Sur Le Traitement du Signal et Ses Applications, Nice, France

Marchisio GB, Hodgkiss WS (1982) Deconvolution applied to a near-bottom seismic profiler. J Acoust Soc Am 72(5):1478-1491

Markel JD (1972) Digital inverse filtering - a new tool for formant trajectory estimation. IEEE Trans Audio Electroacoust 20:129-137

Marucci R (1982) Constrained iterative deconvolution using a conjugate gradient algorithm. Presented at 1982 ICASSP, Paris, France:1845-1857

Mendel JM (1977) A quantitative evaluation of Ott and Meder's prediction error filter. Geophys Prosp 25:692-698

Mendel JM (1977) White noise estimators for seismic data processing in oil exploration. IEEE Trans Auto Cont AC-22:694-706

Mendel JM   (1981)   Minimum-variance   deconvolution.   IEEE   Trans Geosci Rem Sens GE-19:161-171

Mendel   JM   (1982)   Maximum-likelihood   as   applied   to   seismic   inversion problems.   Presented at 6th   IFAC Symposium on Identification and System Parameter Estimation, Washington DC

Mendel JM   (1983)   Inversion   and   deconvolution: foreplay   and   interplay. Invited   tutorial   paper   presented   at   1983   IEEE Intl Symp on Circuits and Systems, Newport Beach CA

Mendel   JM   (983)   Minimum-variance   and   maximum-likelihood   recursive waveshaping.   IEEE Trans Acoust Speech Sig Proc ASSP-31:599-604

Mendel JM   (1983) *Optimal Seismic Deconvolution: An Estimation-Based Approach.* New York: Academic Press

Mendel JM   (1984) Simultaneous correction for divergence and deconvolution without changing industrial practice. Geophysics 49:584-585

Mendel JM (1985) How to include prespecified horizons into minimum-variance deconvolution   and   maximum-likelihood   deconvolution.   Geophysics 50: 1510-1512

Mendel JM (1987) *Lessons in Digital Estimation Theory.* New Jersey: Prentice-Hall

Mendel JM,   Goutsias J   (1986) One-dimensional normal incidence inversion: a solution procedure for bandlimited and noisy data. IEEE Proc 74:401-414

Mendel JM, Kormylo J (1977)   New   fast   optimal   white-noise   estimators   for deconvolution. Special Issue on Geophysical Data Processing of IEEE Trans Geosc Electronics GE-25:32-41

Mendel JM, Kormylo K   (1978)   Single   channel   white-noise   estimators   for deconvolution. Geophysics 43:102-124

Mereu RF (1975) Exact wave-shaping with a time-domain digital filter of finite length.   Symposium   on   Mathematical   Geophysics,   75th Congress of the Canadian Association of Physicists, Toronto, Ontario, Canada

Morley L,   Claerbout J   (1983) Predictive deconvolution in shot-receiver space. Geophysics 48:515-531

Nahman NS,   Guillaume ME (1981) Deconvolution of time domain waveforms in the presence of noise. National Bureau of Standards, NBS Tech Note 1047

Navarro-Guerrero JJ,   Casares-Giner V   (1986)   Systolic   implementation   for deconvolution   iterative   algorithm.   Presented   at   1986   ICASSP,   Tokyo, Japan:1165-1168

Newman BJ (1986) Deconvolution of noisy seismic data -- a method for prestack wavelet extraction. Geophysics 51:34-44

Oldenburg DW   (1981)   A   comprehensive solution to the linear deconvolution problem. J Royal Astr Soc Geophys 65:331-357

Oldenburg DW,   Levy S,   Whittall KP   (1981)   Wavelet   estimation   and deconvolution. Geophysics 46:1528-1542

Omnes G   (1984)   Deconvolution of surface seismic traces using VSP data. In: Toksoz MN   and   Stewart RR   (eds) *Vertical Seismic Profiling, Part B.* London: Geophysical Press, pp. 113-121.

Ooe M,   Ulrych TJ (1979) Minimum entropy deconvolution with an exponential transformation. Geophys Prosp 27:458-473

Oppenheim   AV,   Kopec GE,   Tribolet JM   (1976)   Signal   analysis   by homomorphic prediction.   IEEE Trans Acoust Speech Sig Proc ASSP-24:327-332

Otis RM,   Smith RB   (1977)   Homomorphic   deconvolution by log   spectral averaging. Geophysics 42:1146-1157

Ott N,   Meder HG   (1972) The Kalman filter as a prediction error filter. Geophys Prosp   20:549-560

220   Bibliography

Ozdemir H (1985) Maximum likelihood estimation of seismic reflection coefficients. Geophys Prosp 33:828-860
Pestiaux PM, Berkhout AJ (1986) Spectral analysis by AR and ARMA modeling with application to wavelet estimation for optimal seismic deconvolution. Presented at 48th EAEG Meeting, Ostend, Belgium
Popoli RF, Mendel JM (1989) Heuristically constrained estimation for intelligent signal processing. Presented at the 1987 IEEE Conf on Decision and Control, Los Angeles CA; also, in: Aminzadeh F, Simaan M (eds) Artificial Intelligence and Expert Systems in Petroleum Exploration. Massachusetts: JAI Press
Potts M, Schleicher K, Wason C, Ellender S (1982) Pre-stack wavelet deconvolution. Presented at the 52nd Annual Intl Mtg of the SEG, Dallas TX
Prasad S, Mahalanabis AK (1980) Adaptive filter structures for deconvolution of seismic signals. IEEE Trans Geosci Rem Sens GE-18:267-273
Reimer RB (1973) Deconvolution of seismic response for linear systems. Report No EERC 73-10, Earthquake Engineering Research Center, College of Engineering, Univ of California, Berkeley CA
Rice RB (1962) Inverse convolution filters. Geophysics 2:4-18
Riley DC, Burg JP (1972) Time and space adaptive deconvolution filters. Presented at 42nd Annual Intl Mtg of the SEG, Anaheim CA
Ristow D, Kosbahn B (1976) Time-varying prediction filtering by means of updating. Presented at 38th Mtg of the EAEG, The Hague
Roberts GA, Goulty NR (1988) Directional deconvolution of the seismic source signature combined with prestack migration. First Break 6:247-253
Robinson EA (1954) Predictive decomposition of time series with applications to seismic exploration. PhD Thesis, Mass Inst of Technology, Cambridge MA
Robinson EA (1957) Predictive decomposition of seismic traces. Geophysics 22:767-778
Robinson EA (1962) Random Wavelets and Cybernetic Systems. London: Charles Griffin and Co Ltd
Robinson EA (1963) Mathematical development of discrete filters for the detection of nuclear explosions. J Geophys Res 68:5559-5567
Robinson EA (1967) Multichannel Time Series Analysis with Digital Computer Programs. San Francisco: Holden-Day Inc
Robinson EA (1975) Dynamic predictive deconvolution. Geophys Prosp 23:78-798
Robinson EA (1984) Seismic Inversion and Deconvolution, Part A: Classical Methods, Volume 4A. London, England: Geophysical Press
Robinson EA, Treitel S (1967) Principles of digital Wiener filtering. Geophys Prosp 15:311-333
Robinson EA, Treitel S (1980) Geophysical Signal Analysis. New Jersey: Prentice-Hall
Rocca F, Kostov C (1986) Estimation of residual wavelets. Politecnico di Milano, Italy, private communication
Rocca F, Kostov C (1987) Estimation of residual wavelets. In: Worthington MH (ed) Deconvolution and Inversion. Oxford, England: Blackwell Scientific Publ, pp. 126-144
Romijn R, de Voogd N. Statistical deconvolution of non-white multi layer data. Presented at 46th Meeting of EAEG, London England
Rousseaux P (1986) Deconvolution of time-varying systems by Kalman filtering: its application to the computation of the active state in the muscle. Signal Processing 10:291-301

Saksena BD, Agarwal KC, Pahwa DR, Pradhan MM (1968) Convolution and deconvolution of Lorentzian bands. Spectrochim Acta Part A 24:1981

Sarwar AKM, Smith DL (1987) Wave scattering deconvolution by seismic inversion. Geophys Prosp 35:491-501

Scargle JD (1977) Absolute value optimization to estimate phase properties of stochastic time series. IEEE Trans Info Theory IT-23:140-143

Scheuer TE, Wagner DE (1985) Deconvolution by autocepstral windowing. Geophysics 50:1533-1540

Sengbush RL, Hato M, Chang H (1987) Optimal compensation of time variance and non-minimum phase in Weiner-Robinson deconvolution. In: Worthington MH (ed) *Deconvolution and Inversion*. Oxford, England: Blackwell Scientific Publ, pp. 90-114

Senmoto S, Childers DG (1972) Adaptive decomposition of a composite signal of identical unknown wavelets in noise. IEEE Trans Sys Man Cyber SMC-2:59-66

Senmoto S, Childers DG (1972) Signal resolution via digital inverse filtering. IEEE Trans Aero Electro Sys 8:633-640

Shensa MJ (1976) Complex exponential weighting applied to homomorphic deconvolution. J Royal Astron Soc Geophys 44:379-387.

Shiva M (1982) Geo-optimal deconvolution. PhD Dissertation, Department of Electrical Engineering, Univ of Southern California, Los Angeles CA

Shiva M, Mendel JM (1983) Normal-incidence geo-optimal deconvolution. Presented at the 1983 Intl Geophys Rem Sens Symp, San Francisco CA

Sicking CJ (1980) Windowing and estimation variance in deconvolution. Presented at the 50th Annual Intl Mtg of the SEG, Houston TX

Silvia MT, Robinson EA (1979) *Deconvolution of Geophysical Time Series in the Exploration for Oil and Natural Gas*. Amsterdam: Elsevier

Sims CS, D'Mellow MR (1978) Adaptive deconvolution of seismic signals. IEEE Trans Geosci Elec GE-16:99-103

Stockham TG, Cannon TM, Ingebretsen RB (1975) Blind deconvolution through digital signal processing. IEEE Proc 63:678-692

Stoffa PL, Buhl P, Bryan GM (1974) The application of homomorphic deconvolution to shallow-water marine seismology. Geophysics 39:401-416

Stone DG (1976) Robust wavelet estimation by structural deconvolution. Presented at 46th Annual Intl Mtg of the SEG, Houston TX

Stone DG (1979) Decomposition of seismic traces. Presented at 49th Annual Intl Mtg of the SEG, New Orleans LA

Taylor HL, Banks SC, McCoy JF (1979) Deconvolution with the l_1 norm. Geophysics 44:39-52

Thomas G (1980) Application of the optimal control theory to the deconvolution problem. Presented at 1980 ICASSP, Denver CO, pp. 947-949

Thomas G (1980) Deconvolution and linear tracking problem. Signal Processing 2:131-135

Treitel S, Robinson EA (1964) The stability of digital filters. IEEE Trans Geosci Elec GE-4:25-38

Treitel S, Robinson EA (1966) The design of high resolution digital filters. IEEE Trans Geosci Elec GE-4:25-38

Treitel S, Robinson EA (1977) Deconvolution - homomorphic or predictive? IEEE Trans Geosci Elec GE-15:11-13

Tribolet JM (1979) *Seismic Applications of Homomorphic Signal Processing*. New Jersey: Prentice-Hall Inc

Tribolet JM, Oppenheim AV, Kopec GE (1975) Deconvolution by homomorphic prediction. Presented at 45th Annual Intl Mtg of the SEG, Denver CO

Tugnait JK (1985) Constrained signal restoration via iterated extended Kalman
    filtering. IEEE Trans Acoust Speech Sig Proc ASSP-33:472-475
Ulrych TJ (1971) Application of homomorphic deconvolution to seismology.
    Geophysics 36(4):650-660
Ulrych TJ (1972) Homomorphic deconvolution of some teleseismic events.
    Seismol Soc Am Bull 62:1253-1265
Ulrych TJ, Matsuoka T (1987) The output of predictive deconvolution. Butsuri-
    Tansa 40:274-281
Ulrych TJ, Walkers C (1982) Analytic minimum entropy deconvolution.
    Geophysics 47:1295-1302
Ulrych TJ, Jensen OG, Ellis RM, Sommerville PG (1972) Homomorphic
    deconvolution of some teleseismic events. Seismol Soc Am Bull 62:1269-1281
Ulrych TJ, Smylie DE, Jenson OG, Clarke GKC (1973) Predictive filtering and
    smoothing of short records by using maximum entropy. J Geophys Res
    78:4959-4964
Ursin B, Holberg O (1984) Maximum-likelihood estimation of seismic impulse
    responses. The Foundation of Scientific and Industrial Research at the
    University of Trondheim, SINTEF Report No STF28 A84003, Trondheim,
    Norway
Ursin B, Zheng Y (984) Identification of seismic reflections using singular
    value decomposition. The Foundation of Scientific and Industrial Research at
    the University of Trondheim, SINTEF Report No STFG28 A84004,
    Trondheim, Norway
van Reil P, Berkhout AJ, Poggiagliolmi E (1981) Deconvolution by means of
    parameter estimation. Presented at 51st Annual Intl Mtg of the SEG, Los
    Angeles CA
Walden AT (1985) Non-Gaussian reflectivity, entropy and deconvolution.
    Geophysics 50:2862-2888
Walden AT (1988) A comparison of stochastic gradient and minimum entropy
    deconvolution algorithms. Signal Processing 15:203-211
Walden AT, Hosken JWJ (1986) The nature of the non-Gaussianity of primary
    reflection coefficients and its significance for deconvolution. Geophys
    Prosp 34:1038-1066
Walden AT, Nunn KR (1985) Correcting for coloured primary reflectivity in
    deconvolution. Presented at 47th Meeting of EAEG, Budapest, Hungary
Wang L-X, Dai G-Z (1988) New recursive smoothing algorithms for Bernoulli-
    Gaussian input sequence. Presented at the Intl Fed Auto Cont, Beijing,
    People's Republic of China
Wang RJ (1969) The determination of optimum gate lengths for time-varying
    Wiener filtering. Geophysics 34: 683-695
Wang RJ (1977) Adaptive predictive deconvolution of seismic data. Geophys
    Prosp 25:342-381
Wason CB, Ong CY (1987) Source wavelet estimation by inversion based on
    Laplace transform models. In: Worthington MH (ed) *Deconvolution and
    Inversion*. Oxford, England: Blackwell Scientific Publ, pp. 115-125
Webster GM (ed) (1978) *Deconvolution*. Tulsa, OK: Vols I and II, Geophysics
    Reprint Series, Society of Exploration Geophysicsts
Wertheim GK (1975) Deconvolution and smoothing: applications in ESCA. J
    Electron Spectro Rel Phen 6:239
White RE (1987) Estimation problems in seismic deconvolution. In:
    Worthington MH (ed) *Deconvolution and Inversion*. Oxford, England:
    Blackwell Scientific Publ, pp. 5-37

Wiggins RA (1977) Minimum entropy deconvolution. Proc Intl Symp Computer Aided Seismic Analysis and Discrimination, Falmouth, MA; IEEE Computer Soc:7-14

Wiggins RA (1978) Minimum entropy deconvolution. Geoexploration 16:21-35

Wiggins RA (1985) Entropy guided deconvolution. Geophysics 50:2720-2726

Wiggins RA, Robinson EA (1965) Recursive solution to the multichannel filtering problem. J Geophys Res 70:1885-1891

Wu Z-X (1986) A new homomorphic deconvolution system. Presented at 1986 ICASSP, Tokyo, Japan

Yanagida M, Kakusho O (1982) Least-squares method for multi-dimensional deconvolution. Presented at 1982 ICASSP, Paris, France:1849-1852

Yarlagadda R (1985) Fast algorithms for l_p deconvolution. IEEE Trans Acoust Speech Sig Proc ASSP-33:174-182

Yeung W-K (1986) Time domain deconvolution when the kernel has no spectral inverse. IEEE Trans on Acoust Speech Sig Proc ASSP-34:912-918

Yu K-B, Sistanizadeh MK (1986) Complex deconvolution in non-coherent radar systems. Presented at 1986 ICASSP, Tokyo, Japan:1921-1924

Zhao X, Mendel JM (1988a) Minimum-variance deconvolution using artificial neural networks. Presented at the 58th Annual Intl Mtg of the SEG, Anaheim CA

Zhao X, Mendel JM (1988b) An artificial neural minimum-variance estimator. Presented at 1988 IEEE Intl Conf on Neural Networks, San Diego CA

Ziolkowski AM (1982) Further thoughts on Popperian geophysics -- the example of deconvolution. Geophys Prosp 30:155-165

Ziolkowski AM (1984) *Deconvolution*. Massachusetts: Intl Human Res Dev Corp Publishers

Ziolkowski AM, Lerwill WE, March DW, Peardon LG (1980) Wavelet deconvolution using a source scaling law. Geophys Prosp 28:872-901

Index